Innovation and Discovery in Russian Science and Engineering

This Series provides rapid dissemination of the most recent and advanced work in engineering, science, and technology originating within the foremost Russian Institutions, including the new Federal District Universities. It publishes outstanding, high-level pure and applied fields of science and all disciplines of engineering. All volumes in the Series are published in English and available to the international community. Whereas research into scientific problems and engineering challenges within Russia has, historically, developed along different lines than in Europe and North America. It has yielded similarly remarkable achievements utilizing different tools and methodologies than those used in the West. Availability of these contributions in English opens new research perspectives to members of the scientific and engineering community across the world and promotes dialogue at an international level around the important work of the Russian colleagues. The broad range of topics examined in the Series represent highly original research contributions and important technologic best practices developed in Russia and rigorously reviewed by peers across the international scientific community.

More information about this series at http://www.springer.com/series/15790

Alexander A. Lyapin • Ivan A. Parinov
Nina I. Buravchuk • Alexander V. Cherpakov
Ol'ga V. Shilyaeva • Ol'ga V. Guryanova

Improving Road Pavement Characteristics

Applications of Industrial Waste and Finite Element Modelling

Springer

Alexander A. Lyapin
Don State Technical University
Rostov-on-Don, Russia

Nina I. Buravchuk
I. I. Vorovich Institute of Mathematics,
Mechanics and Computer Science
Southern Federal University
Rostov-on-Don, Russia

Ol'ga V. Shilyaeva
Don State Technical University
Rostov-on-Don, Russia

Ivan A. Parinov
I. I. Vorovich Institute of Mathematics,
Mechanics and Computer Science
Southern Federal University
Rostov-on-Don, Russia

Alexander V. Cherpakov
I. I. Vorovich Institute of Mathematics,
Mechanics and Computer Science
Southern Federal University
Rostov-on-Don, Russia

Ol'ga V. Guryanova
I. I. Vorovich Institute of Mathematics,
Mechanics and Computer Science
Southern Federal University
Rostov-on-Don, Russia

ISSN 2520-8047 ISSN 2520-8055 (electronic)
Innovation and Discovery in Russian Science and Engineering
ISBN 978-3-030-59229-5 ISBN 978-3-030-59230-1 (eBook)
https://doi.org/10.1007/978-3-030-59230-1

This Springer imprint is published by the registered company Springer Nature Switzerland AG
The registered company address is: Gewerbestrasse 11, 6330 Cham, Switzerland

Centenary dedicated to academician Iosiph Izrailevich Vorovich

Preface

In the context of increasing demand for natural raw materials, the problem of the integrated use of all extracted raw materials and environmental protection is becoming of paramount importance and is more and more acute every year. The mineral resource base is characterized by a further decrease in quality and reserves and also a complication of the conditions for developing mineral deposits. Moreover, even with modern technology for the extraction and processing of mineral raw materials, only about 5% of mineral resources are used. The bulk goes into technogenic waste. As a result, huge stocks of waste from the extraction and processing of natural raw materials are stored in dumps on the earth's surface. Therefore, integrated development of natural mineral resources must involve the mandatory processing of waste into useful products as the final stage in the extraction of minerals. The rate and scale of anthropogenic impact on the environment in some cases exceed the potential of the biosphere for self-regulation. Huge reserves of burnt mine rocks and ash-slag waste create environmental problems at their locations. Due to this, the urgent task is the practical implementation of the concept of sustainable (self-sustaining) development.

Road construction is the most material-intensive industry for the use of nonmetallic building materials. The increasing rates of road constructions and the need to reduce material and financial resources and fill the shortage of natural raw materials are directing to the widespread use of cheaper local materials and technogenic waste. Modern scientific and technological progress allows one to utilize large-tonnage production of waste with a significant economic effect due to the replacement of natural types of raw materials in a large scale. Not only the health and well-being of the current and future generation of people but also the development of civilization and the existence of all mankind depend on the successful resolution of these issues.

Russia, with tremendous mining, has accumulated a great experience in processing technogenic waste [1–3]. This book presents some latest achievements and results in resource-saving area. Moreover, a significant attention is devoted to theoretical approaches and numerical methods for studying stress-strain state of road constructions on various soils, taking into account their structural geometry and physico-mechanical parameters of materials, which are constructed with widely used technogenic waste.

The book consists of two parts and is divided into ten chapters. Part I includes first seven chapters and discusses resource-saving technologies and technogenic waste. Part II includes last three chapters and is devoted to mathematical modeling of road constructions on different soils.

Chapter 1 describes the problems of using technogenic raw materials in Russia including the reasons and factors which hinder the application of the mass technogenic waste in the national industry. Moreover, the necessary ways allowing increasing use of mine rocks and ash-slag waste are discussed.

Chapter 2 discusses main areas of utilizing technogenic waste (burnt rocks of mine dumps and ash-slags). With this aim, there are stated assessment criteria for multilevel testing of waste according to the environmental requirements, chemical and mineral compositions, hydraulic activity, physical and mechanical properties, structural uniformity, formed volume in dumps, and economic indicators of their application. Based on certain examples, Russian and world experiences in the utilization of burnt rocks of mine dumps and ash-slags from the coal industry and heat power industry are considered.

Ways of improving the structural properties of burnt rocks and ash-slags are present in Chap. 3. First, technological recycling schemes of complex enrichment of rock materials with various strength are discussed taking into account different stages of technological process. In particular, the technological operations of crushing of raw materials, sorting of crushing products, and transportation and storage of finished products are considered. Equipment for crushing and screening complex is discussed in some details. Main characteristics of crushed stone quality are studied at each technological stage. Then the results of characterization of ash-slag mixtures used in road construction are present. The obtained data are based on the practical experience of studying burnt rocks, fly ashes, and slags from dumps of several Russian mines.

Chapter 4 treats composition and properties of the burnt rocks of mine dumps and ash-slag waste. Beginning with formation sampling, physical and chemical properties and granulometric and mineralogical compositions of samples from different Russian mines are studied by physico-mechanical methods. Then hydraulic activity of ash and burnt rocks of mine dumps, allowing one to use these wastes as active mineral additives, is studied. Kinetics of the involvement of additives in the reaction with calcium hydroxide are discussed. Changes in the specific surface of the additive composition and in the adsorption activity of a mixture of fly ash and burnt rocks are considered.

Chapter 5 discusses the use of the burnt rocks of mine dumps and ash-slag waste for the building of subgrade road foundations, and as fillers of these products in the constructional layers of pavements on the base of practical experience accumulated in the Rostov region (Russia). For each constituent of road construction, certain materials are treated, which can allow attainment of the optimal properties and long-life service of road pavement. In particular, hardening kinetics of compositions containing technogenic raw materials are studied and also application of these materials in asphalt concrete and concrete coatings. The main characteristics, defining strength and durability of the constructions, are considered as follows: water saturation, water resistance, frost resistance, adhesion of bitumen to the mineral, abrasion, cement saving, etc. Moreover, replacement of ordinary concrete by the concrete products based on mineral fillers and binders is discussed. Compression strength, bending strength, and kinetics of the development of linear strains of concrete during cyclic tests in the conditions of aggressive medium are studied.

Physicochemical fundamentals of hardening the burnt rocks and ash-slag waste used in road pavements are presented in Chap. 6. The discussion is carried out for different constituents of road construction (subgrade, base, and structural layers of pavement). The effects of fillers from burnt rocks and ash-slag waste on the structure formation of asphalt concrete and concrete are also considered with estimation of ways to attain high strength of such structures. In particular, among the hardening products, neoplasms formed during hydration and hardening of clinker minerals are noted. Moreover, thermal properties of hydration products are studied during hardening of concrete.

Chapter 7 is devoted to ecological and economic assessments of the efficiency of using the burnt rocks of mine dumps and ash-slag waste in road constructions. These estimations are based on the prevented damage to environment and also on the assessment of the efficiency of technogenic waste use as an industrial subsector being an element of the regional economic complex system directed to complete satisfaction of the social economic needs of the region.

Dynamic modeling of solid mass on soil base is in the center of Chap. 8. Features of finite-element modeling soil massif are discussed on the base optimal selection of representative volume with damping belts. As the test task, it is considered lattice of partitioning a three-layer half-plane into finite elements, with the introduction of damping belts and taking into account a coastal slope. The dynamic stress-strain states of these structures are studied for different geometrical parameters of the layers and their mechanical characteristics. Finally, comparison of finite-element calculations is performed with analytical solutions, obtained by using an integral method of harmonic analysis based on the application of the Fourier transform in time.

Based on numerical experiments, Chap. 9 deals with the basic laws of distribution of amplitude characteristics of oscillations propagating in soils of various structures in the presence and absence of a coastal slope. The consideration distinguishes two cases of the ratio of the stiffnesses of the layers of the investigated structure: (i) the normal structure, when the stiffnesses of the layers increase with the depth of the layered structure, and (ii) the anomalous structure in which the pointed regularity is violated. Two types of anomalous structures with a distribution of layer stiffnesses from top to bottom are considered: (i) "hard–soft–hard" and (ii) "soft–hard–soft." Moreover, these numerical results are obtained taking into account that an oscillation load is located to the left or right of the slope.

Chapter 10 presents the results of a study of the dynamic deformation characteristics of elements of layered extended structures using the example of the "road construction–soil" system. The FEM model developed and the results obtained on its base are used to modernize methods for predicting the process of damage accumulation in the elements of road constructions. For modeling semi-infinite media, the plane, spatial, and symmetric models are used. Deflections of the surface of coating are studied depending on geometric, physical, and mechanical characteristics of the considered layered structures.

This self-sufficient book, presenting resource-saving technologies, based on the use of technogenic raw materials and their application for development of road pavements, together with finite-element modeling of soils and multilayer structures, is intended to a wide range of students, engineers, and specialists interested and participating in R&D of the modern problems of the mechanics of solids, building constructions, and ecology.[1]

Rostov-on-Don, Russia
May 2020

Alexander A. Lyapin
Ivan A. Parinov
Nina I. Buravchuk
Alexander V. Cherpakov
Ol'ga V. Shilyaeva
Ol'ga V. Guryanova

References

1. Trubetskoy, K. N., & Umanets, V. N. (1992). *Gorny Journal, 1*, 12. (In Russian).
2. Trubetskoy, K. N., Umanets, V. N., & Nikitin, M. B. (1987). *Gorny Journal, 12*, 18. (In Russian).
3. Trubetskoy, K. N., Umanets, V. N., & Nikitin, M. B. (1989). *Gorny Journal, 12*, 6. (In Russian).

[1] Research was financially supported by Southern Federal University, grant No. VnGr/2020-04-IM (Ministry of Science and Higher Education of the Russian Federation).

Contents

Part I Resource-saving Technologies

1 Problems of Using Technogenic Raw Materials 3

2 Main Areas of Utilizing the Burnt Rocks of Mine Dumps and Ash-slag Waste 9

2.1 Comprehensive Assessment of Technogenic Raw Materials 9

2.1.1 First Level: Assessment of Environmental Indicators 10

2.1.2 Second Level: Assessment of Chemical Composition 11

2.1.3 Third Level: Assessment of Mineral Composition 11

2.1.4 Fourth Level: Assessment of Reactivity (Activity) 11

2.1.5 Fifth Level: Assessment of Physical and Mechanical Properties 12

2.1.6 Sixth Level: Assessment of Uniformity 12

2.1.7 Seventh Level: Assessment of Formed Volume 12

2.1.8 Eighth Level: Assessment of Technical and Economic Indicators 12

2.2 Burnt Rocks of Mine Dumps 13

2.3 Ash-slag Waste 16

3 Ways of Improving the Structural Properties of Burnt Rocks of Mine Dumps and Ash-slag Waste 19

3.1 Burnt Rocks of Mine Dumps Recycling Schemes 19

3.2 Non-ore Building Materials from the Burnt Rocks of Mine Dumps and Ash-slag Waste 33

3.2.1 Properties of Burnt Aggregates from Mine Dumps 33

3.2.2 Properties of Ash-slag Aggregates 37

4 Composition and Properties of the Burnt Rocks of Mine Dumps and Ash-slag Waste 41

4.1 Sampling and Research Methods 41

4.2 Burnt Rocks of Mine Dumps 42

4.2.1 Formation of the Burnt Rocks of Mine Dumps 42

4.2.2 Physical Properties of the Burnt Rocks of Mine Dumps . . . 44
4.2.3 Chemical Composition of the Burnt Rocks of Mine Dumps . . . 45
4.2.4 Mineralogical Composition of the Burnt Rocks of Mine Dumps . . . 48
4.3 Ash-slag Waste . . . 59
4.3.1 Methods of Ash-slag Waste Selection . . . 59
4.3.2 Physical Properties of Ash-slag Waste . . . 60
4.3.3 Chemical Composition of Ash-slag Waste . . . 63
4.3.4 Mineralogical Composition of Ash-slag Waste . . . 64
4.4 Hydraulic Activity of Ash and Burnt Mine Dumps . . . 69

5 The Use of Burnt Rocks of Mine Dumps and Ash-slag Waste in Road Construction . . . 77
5.1 Subgrade Composition . . . 77
5.2 Materials from Technogenic Raw Materials for Pavement Base Layers . . . 84
5.3 Structural Layers of Road Pavements from Burnt Rocks of Mine Dumps . . . 90
5.4 Asphalt Concrete Coatings Using Technogenic Raw Materials . . . 97
5.5 Concrete Coatings Using Technogenic Raw Materials . . . 103

6 Physicochemical Fundamentals of Hardening the Burnt Rocks of Mine Dumps and Ash-slag Waste in Road Pavements . . . 113
6.1 Subgrade Hardening . . . 113
6.2 Increase of Strength in the Bases and Structural Layers of Pavement from Technogenic Raw Materials . . . 114
6.3 The Formation of the Structure of the Asphalt Concrete Composition Based on Technogenic Raw Materials . . . 115
6.4 The Effect of Materials from Burnt Rocks and Ash-slag Waste on the Structure Formation of Concrete . . . 116

7 On the Efficiency of Using the Burnt Rocks of Mine Dumps and Ash-slag Waste in Road Constructions . . . 123
7.1 Ecological and Economic Assessment of the Utilization of the Burnt Rocks of Mine Dumps . . . 123
7.2 Efficiency of Application of Fly Ash and Burnt Ash in Building Materials . . . 126
7.3 Concluding Remarks . . . 129

Part II Finite-element Modeling of Road Constructions on Soils

8 Dynamic Modeling of Solid Mass on Soil Base . . . 133
8.1 Dynamics of "Solid Mass-Soil" System . . . 133
8.2 Features of Finite-Element Modeling Soil Massif . . . 134

8.3 Finite-Element Model of System: Test Tasks 137
8.4 Test Calculations and Analysis of Results 138
8.5 Comparison of Finite-Element Calculations with Analytical Solutions 145

9 Studying Characteristics of Waves Propagating in Layered Structure with Semi-infinite Layers 151
9.1 Normal Structure (Stiffness of Layers Increases with Depth) 152
9.1.1 Source of Oscillations Disposes to the Left of Slope 155
9.1.2 Source of Oscillations Disposes to the Right of Slope 161
9.2 Anomalous Structure (Stiffness of Layers Changes with Depth Not Monotonously) 171
9.2.1 Distribution of Layer Stiffness in Depth: "Hard–Soft–Hard" with Oscillation Source Disposed to the Left or Right of Slope 172
9.2.2 Distribution of Layer Stiffness in Depth: "Soft–Hard–Soft" with Oscillation Source Located to the Left or Right of Slope 182

10 Modeling Pavement Constructions 189
10.1 Statement of Problems into Framework of FEM 190
10.2 Features of Modeling Semi-infinite Media 193
10.2.1 Spatial Model 197
10.2.2 Symmetric Model 198
10.2.3 Plane Model 200
10.3 Comparative Analysis of Characteristics 200
10.3.1 Initial Data of Numerical Experiment 201
10.3.2 Response of the Surface of Coating to Impact Loading for Different Mechanical Properties of Road Construction Elements 202
10.3.3 Features of Deforming the Various Road Constructions 204
10.4 Concluding Remarks 222

References 223

Index 231

Contributors

Nina I. Buravchuk I. I. Vorovich Institute of Mathematics, Mechanics and Computer Science, Southern Federal University, Rostov-on-Don, Russia

Alexander V. Cherpakov I. I. Vorovich Institute of Mathematics, Mechanics and Computer Science, Southern Federal University, Rostov-on-Don, Russia

Ol'ga V. Guryanova I. I. Vorovich Institute of Mathematics, Mechanics and Computer Science, Southern Federal University, Rostov-on-Don, Russia

Alexander A. Lyapin Don State Technical University, Rostov-on-Don, Russia

Ivan A. Parinov I. I. Vorovich Institute of Mathematics, Mechanics and Computer Science, Southern Federal University, Rostov-on-Don, Russia

Ol'ga V. Shilyaeva Don State Technical University, Rostov-on-Don, Russia

Part I
Resource-saving Technologies

Chapter 1
Problems of Using Technogenic Raw Materials

Abstract Introductory Chap. 1 describes the problems of using technogenic raw materials in Russia including the reasons and factors which hinder the application of the mass technogenic waste in the national industry. Moreover, the necessary ways allowing increasing use of mine rocks and ash-slag waste are discussed.

Keywords Technogenic raw materials · Ash-slag waste · Environmental safety · Burnt mine rocks · Deposits

Every year, the reserves of natural raw materials are depleted. At complicated conditions for its production, the cost of natural mineral raw materials increases. The difficulty of finding stone or gravel pits with the required characteristics and that are located close to the consumer increases. Some regions of Russia practically no longer have their own raw material reserves of fillers and are forced to import them from other regions and territories. This situation, as well as rising prices for minerals for road construction and for transportation services to deliver it to the place of consumption, and the need for fees for the use of subsoil and their development, all these causes encourage manufacturers to search for alternative solutions, among which of greatest interest is the use of industrial waste [44, 50, 61, 148, 168]. The prospect of using technogenic raw materials is obvious. The concept of integrated development of natural mineral resources involves the introduction of low-waste technologies and mandatory waste processing into useful products as the final stage in the extraction of minerals. Such an approach to the use of natural resources will ensure the environmental safety of the country and the preservation of its natural resource potential. The task of improving the environmental situation in industrialized regions and expanding the range of new materials makes this stage an indispensable necessity. First of all, this concerns large-scale waste, which includes associated products of the power system and coal mining waste. In the coal industrial regions of any country, huge reserves of mine rocks and ash-slag waste are concentrated in dumps. The maintenance and storage of the dumps of these

A. A. Lyapin et al., *Improving Road Pavement Characteristics*, Innovation and Discovery in Russian Science and Engineering,
https://doi.org/10.1007/978-3-030-59230-1_1

products are quite expensive. Their dusty fractions, containing gaseous compounds of oxides of sulfur, carbon, nitrogen, and other substances, significantly affect the formation of the ecological situation in the areas of their formation. This issue of wastes has been discussed since many years, but no particular success has been achieved in solving the problem. The volumes of utilization of coal mining and burning waste are 5–6%, although the potential opportunities are much wider, and the volume of utilization can be brought up to 60% and even more. Available scientific and technical developments allow us to consider coal mining and heat energy waste as an important source of raw materials for road construction.

The works of many researchers have been devoted to solving the scientific and technical problems of the integrated development of technogenic deposits [90, 107, 117, 132, 133, 159, 162]. With significant volumes of rocks from mine dumps and ash-slag waste, their level of disposal is low. When analyzing the reasons that hinder the mass involvement of technogenic raw materials in the national economy, the following factors are usually distinguished:

(i) Instability of the properties of technogenic raw materials; insufficient knowledge of the material composition and technological properties; lack of a unified classification
(ii) Imperfection in the pricing of mineral and produced raw materials
(iii) Interdepartmental fragmentation of enterprises performing the extraction, enrichment, and use of mineral and technogenic raw materials
(iv) Insufficient economic interest of enterprises and departments in the use of technogenic raw materials, a great prejudice of production workers towards waste as raw materials
(v) Lack of funding for additional capital construction for the processing of technogenic raw materials
(vi) Difficulties in the reconstruction of existing enterprises for the production of goods using technogenic raw materials
(vii) Lack of information on the environmental safety of waste
(viii) Lack of regulatory and technical documentation for the use of waste
(ix) Incomplete completion and insufficient technical implementation of technologies for enrichment, processing, and use of technogenic raw materials
(x) Lack of effective control over the implementation by enterprises of environmental safety requirements for production, extraction, enrichment, processing, and use of natural resources
(xi) Lack of interest of owners of ash dumps in the processing of this valuable source of raw materials
(xii) Limited opportunities for shipment and transportation of industrial raw materials

For decades, Russian national economy has focused on the use of natural resources. This restricts the development and use of industrial waste. The low cost, availability for development, huge volumes of existing reserves, and significant volumes of their annual formation and accumulation testify in favor of processing and using technogenic raw materials.

One of the main reasons restraining the disposal of mine rocks and ash-slag waste is the instability of their properties. They have significant heterogeneity in chemical and mineralogical composition, particle size distribution, strength, and resistance to weather and climatic factors, which requires a differentiated approach to these materials. The instability of the properties of mine rocks is associated with the specific formation of waste dumps. Carbon-bearing rocks of various coal seams can be stored in dumps. Due to the variety of properties, composition, conditions of formation, occurrence, and development of carbon-bearing rocks and coal seams, mine rocks of various pools and even within the same basin and even dump are distinguished by their material composition and physico-mechanical properties. In the mine waste dump, there are rocks of various degrees of firing, unburnt, and remelted rocks, different in structure and size.

The properties of ash-slag waste depend on the type and composition of the fuel burnt. The variety of solid fuels, the conditions and mode of combustion, and the method of selection of combustion residues lead to the formation of ash and slag waste with various properties. High levels of unburned fuel may be present in the waste. In each case, it is necessary to evaluate the aggregate strength of the waste, characterizing their construction properties.

The volumes of consumption of burnt mine rocks and ash-slag waste will increase significantly, provided that their properties are averaged and their quality is brought to standard materials. To do this, it is necessary to provide for their processing at rock and ash-slag dumps with the aim of issuing conditioned products to the consumer. Financing is needed to set up waste-processing facilities, including budgetary funds and/or funds of waste owners.

Another way to reduce heterogeneity, for example, mine rocks, is the targeted formation of new technogenic deposits through the selective development and separate storage of simultaneously produced rocks for their subsequent development, when their reprocessing becomes economically feasible. With the purposeful formation of technogenic deposits, it is necessary to develop effective technologies for the storage of waste, and methods and means of their conservation, and to solve transport problems during selective excavation of rocks. The purposeful formation of technogenic deposits increases the cost of developing mineral resources. The use of mathematical modeling and optimization of field development parameters will allow minimizing additional costs. The general principles of the formation of technogenic deposits, the assessment of their prospects, and the development of effective technologies for their integrated use have been the subject of the works of many researchers [11, 12, 13, 52, 92, 141, 142, 143, 144, 145, 146, 150, 152, 160].

The transition to a market economy did not ensure an increase in the rate of waste processing. The need to combine the flexibility of a market economy that is capable of a quick raw material reorientation with government support stimulating the use of waste and reducing its negative impact on the environment has become extremely acute. For example, the processing of ash-slag waste for power engineers is a by-product; it does not have support at the Russian federal level. Each company and each region decide this problem alone, often from time to time. We should move on to mass production and technology, and this requires a state strategy for the use of slag. We need clear standards for the use of slag in construction. Finally, it is necessary to achieve recognition of ash-slag waste not as waste but as goods, which will allow designers to justify the use of these wastes already at the stage of project development. Legislative and national measures are required, which, on the one hand, would force producers to sell industrial wastes, and on the other hand would stimulate their use. The same measures are necessary to increase the use of burnt rock dumps. The issue of benefits for the use of subsoil, depending on the completeness of waste use, should be resolved legislatively. Disposal of large-capacity industrial wastes requires additional material costs, sometimes quite significant, associated with the reorganization of the main production.

To increase the level of use of mine rocks and ash-slag waste, it is necessary:

(i) To adjust the prices of technogenic raw materials and products from it
(ii) To develop a system of economic levers of management that increases the economic interest of enterprises in the use of technogenic deposits
(iii) To introduce incentives to enterprises (both suppliers and consumers of waste) in the form of reduced payments for fixed assets intended for the processing and preparation of waste for consumption
(iv) To reduce transport tariffs for waste transportation
(v) To reduce the value-added tax and profit in proportion to the share of industrial waste used in the production
(vi) To exclude these taxes for the period of modernization of production and the development of advanced technologies and equipment for waste processing

The successful solution of environmental problems and the integrated development of technogenic deposits will be facilitated by the improvement of credit, investment, tax policies, and other economic standards applicable to specific regions.

Moreover, as experts say, another important point is to create a complete information base on the methods and technologies for processing mine rocks and ash-slag waste, implemented and planned projects, and their investment potential. The lack of such information is a real brake on decision-making by business leaders in the processing of mine rocks, ashes, and slags.

For the organization of the integrated development of technogenic deposits, it is very important to have complete information about their composition and properties and possible directions for their utilization and economic evaluation of their processing. In a market economy, the integrated use of mineral resources can only attract economic interest. This information allows us to evaluate the feasibility of building the corresponding waste-processing enterprises within the framework of the created territorial production complex in conjunction with other structural components. In practice, these are new productions aimed at implementing new technological solutions. It is most advisable to create such production complexes in the immediate vicinity of the source of generation and the greatest accumulation of waste, or directly as part of an enrichment or coal-mining enterprise.

The processing may include burnt and unburned rocks, waste from coal preparation, and burning of coal. Burning of mine dumps should also be considered, the development of which can begin in the future. In the structure of the production complex, the central place should be occupied by enterprises for bringing waste to the quality characteristics of traditional raw materials. Waste prepared in this way should become a raw material for production structures that produce on the base of technogenic raw materials additional useful goods that are not typical for this enterprise of industry affiliation.

Thus, the opportunity arises to form large industrial coal-mining enterprises operating on low-waste and non-waste technology with minimal negative impact on the environment. Similar enterprises for processing waste of current output and accumulated in dumps can be created at ash dumps. By fully extracting and using all the valuable components of extracted useful raw materials and that are simultaneously extracted for various needs of the national economy, the principle of integrated development of mineral resources is implemented.

Chapter 2
Main Areas of Utilizing the Burnt Rocks of Mine Dumps and Ash-slag Waste

Abstract This chapter discusses main areas of utilizing technogenic waste (burnt rocks of mine dumps and ash-slags). With this aim, there are stated assessment criteria for multilevel testing of waste according to the environmental requirements, chemical and mineral compositions, hydraulic activity, physical and mechanical properties, structural uniformity, formed volume in dumps, and economic indicators of their application. Based on certain examples, Russian and world experiences in the utilization of burnt rocks of mine dumps and ash-slags from the coal industry and heat power industry are considered.

Keywords Mine dumps · Industrial waste · Waste strengthening technology · Road construction · Cement stone · Chemical composition · Mineral composition · Physico-mechanical properties · Multilevel assessment

2.1 Comprehensive Assessment of Technogenic Raw Materials

In the context of increasing demands for mineral resources, the task of rational nature management is especially acute, allowing one to satisfy the needs of production, not forgetting about the protection and reproduction of the environment. At the same time, the ideology of unlimited technological progress gives way to the concept of sustainable development [112], which takes into account the interests of not only the present but also future generations. One of the directions of this concept is the use of industrial waste accumulating in dumps and representing technogenic raw materials. The calculations of specialists show that in the case of complex use of raw materials and industrial goods, the output of many types of products can be increased by no less than 30%. Resource conservation is becoming comprehensive and is part of the economic worldview.

A. A. Lyapin et al., *Improving Road Pavement Characteristics*, Innovation and Discovery in Russian Science and Engineering,
https://doi.org/10.1007/978-3-030-59230-1_2

To solve the problem of involving technogenic raw materials in the production, it is necessary to conduct a comprehensive assessment of its quality. This problem has attracted many researchers [21, 60, 80, 100, 102, 112, 135, 151]. At the stage of forecasting the possible directions of disposal of mine rocks and ash-slag waste, it is necessary to conduct multilevel testing of waste according to the following criteria.

2.1.1 First Level: Assessment of Environmental Indicators

The ecological purity of the waste is estimated by comparing the composition of the waste with the maximum permissible values MPVs) of heavy metals and toxic substances. The criterion of ecological purity of waste is also the value of the specific effective activity of natural radionuclides. To determine the presence of trace elements, a spectral analysis of the waste is carried out. So, from the analysis of long-term studies of the rocks of mine dumps, it follows that the studied waste contains in its composition ingredients in the amount: 0.006–1.0% (Ft); 0.005–0.02% (Zn); 0.06–0.08% (Mn); 0.003–0.01% (Cu); and 0.008–0.01% (Pb). The content of Ni, Cd, Co, Sn, Bi, and Cr does not exceed 0.006%.

If the content of individual elements of heavy metals is slightly higher than the MPVs, then this is not an obstacle to the disposal of specific technogenic raw materials. By using such raw materials in the technology of building materials when forming the structure of a new material, as a rule, blocking of potentially dangerous elements occurs. Concrete, roasting, and waste strengthening technology in road construction makes it possible to obtain capsules with heavy metals inside the structure of the material. Such capsules not only are durable, but also provide the burial of heavy metals [134]. Special studies were carried out [70] and the ecological and hygienic properties of cement stone samples based on finely ground binders containing ashes of thermal power plants, in which the content of some potentially dangerous elements exceeds the maximum permissible value, were determined. Studies have shown that the emission of harmful substances from samples of cement stone containing industrial wastes is significantly less than from raw materials, and basically does not exceed hygienic standards. Mine waste is a promising raw material for the production of building materials. To make a decision on the recommendations for the disposal of rocks of a specific dump, an express methodology is needed to assess the content of potentially hazardous ingredients in the rocks.

The specific effective activity of natural radionuclides from mine dumps and ash-slag waste, according to the results of numerous radiological studies of special laboratories, did not exceed 370 Bq/kg (it was in the range of 144–280 Bq/kg). In accordance with regulatory requirements, the studied burnt mine rocks in terms of the effective specific activity of radionuclides belong to class I materials, the use of which is unlimited.

2.1.2 Second Level: Assessment of Chemical Composition

The chemical composition of mine rocks and ash-slag waste is represented mainly by oxides of silicon, aluminum, and iron. The content of acid oxides is more than 70% and the waste can be characterized as siliceous. Based on modulo basicity, taking into account the ratio of acid and basic oxides, rocks are acidic. Of significant importance in choosing the direction of disposal of these wastes is the quantitative sulfur content. According to this indicator, the waste is classified as medium sulfur. The presence of an insignificant amount of sulfur compounds does not create complications by using this technogenic raw material. The content of soluble oxides in the composition of the rocks is small.

2.1.3 Third Level: Assessment of Mineral Composition

The mineral composition of technogenic raw materials is significantly affected by the technological features of its formation. Mine rocks and ash-slag waste differ from traditional raw materials in the material composition and structure of the source minerals. The mineralogical and petrographic characteristics of mine rocks and ash-slag waste allow us to identify the presence of rocks classified as harmful impurities, the content of amorphous components, the presence of carbonaceous substances, and their modifications. The content of amorphous species of silicon dioxide, soluble in alkalis, in the studied waste does not exceed 50 mmol/l; mica and layered silicates 15%; and magnetite, hematite, and iron hydroxides 10% by volume. Carbon impurities are present mainly in a modified form. They are represented by coking products (semicoke and coconut residues), colloidal carbon sprayed onto the surface of mineral particles, and graphitized carbonaceous matter. These varieties of carbonaceous substances are resistant to oxidation; their durability when exposed to moisture is quite high. According to the content of the rocks classified as harmful impurities, the studied waste does not exceed the requirements of regulatory documents.

2.1.4 Fourth Level: Assessment of Reactivity (Activity)

Establishing the presence of components that are in a metastable state allows us to evaluate the degree of activity of the waste and recommend it for use as the main raw material or additives. Burnt mine rocks and ash-slag waste are active mineral additives. They are able to show hydraulic activity. Their mineral part is represented by rocks amorphized during high-temperature firing and consisting of a metastable clay substance with the inclusion of a glass phase. Such systems have increased physicochemical activity. In their composition they contain a significant amount of clay, and glandular and siliceous hydraulic components. The presence of these components is associated with a violation of the crystal lattice of clay minerals during self-firing of rocks and burning of solid fuel. Firing products gain some energy potential. This new state of the substance is the reason for the ability of firing products to hydrate.

2.1.5 *Fifth Level: Assessment of Physical and Mechanical Properties*

Technogenic wastes are characterized by strength, abrasion, water resistance, ductility, dispersion, density, and other properties, the indicators of which significantly affect the scope of their application.

2.1.6 *Sixth Level: Assessment of Uniformity*

According to the chemical and mineralogical composition and physico-mechanical properties, mine rocks and ash-slag waste within the same dump are not uniform. Studying the properties of wastes makes it possible to assess the degree of their preparedness for participation in technological processes, and to determine effective ways to bring their characteristics to indicators of quality raw materials. Waste treatment and processing technologies that provide the highest maximum yield of conditioned products are preferred.

2.1.7 *Seventh Level: Assessment of Formed Volume*

The volumes of formation of mine rocks and ash-slag waste allow us to evaluate them as large-tonnage waste. The reserves of rock mass and ash-slag waste are huge, and their accumulations in dumps can be considered as technogenic deposits. They can be used as the main raw material, or corrective additives.

2.1.8 *Eighth Level: Assessment of Technical and Economic Indicators*

When choosing directions for the utilization of mine rocks and ash-slag waste, one should be guided by the following criteria:

(i) Low cost of this raw material in comparison with the full development of natural deposits
(ii) Minimum number of operations for their preparation for use
(iii) The possibility of replacing scarce traditional raw materials
(iv) A small distance of their transportation to the place of consumption
(v) High properties of materials and compositions made on the base of burnt rocks of mine dumps or slags
(vi) Saving fuel and energy resources
(vii) High rates of physical-mechanical and construction-operational properties of materials and compositions obtained on the base of these wastes
(viii) The demand for materials and compositions based on burnt rocks of mine dumps and ash-slag waste

A multilevel assessment allows one to determine the effective directions of the utilization of mine rocks and slag, and to choose effective technological schemes for their preparation and processing. Mine rocks and ash-slag waste are raw materials for filling of the mined-out area; reclamation of land disturbed by mining operations; creation of structural embankments; construction of industrial production sites, parking lots, stopping areas, stadiums, and tennis courts; landscaping; hydraulic engineering; and road constructions. The need for such products for road construction is increasing every day. Roads, providing 80% of all freight and 70% of passenger traffic, are one of the bottlenecks in the transport system of any country. To build, first of all, local access roads, technological roads, city roads, and sidewalks, a huge amount of nonmetallic building materials will be required. This niche can be filled by materials from mine rocks and slag. These materials can be used in the construction, reconstruction, and repair of roads of regional and local significance, and a large network of rural roads. These materials will be very effective in the implementation of the program for the reconstruction of soil roads. The practical experience of processing burnt mine rocks and the study of the physico-mechanical properties of the fillers obtained from them [20, 22, 26, 32, 38] confirm the applicability of these materials for arranging structural layers of pavement and other general construction works. Examples of the use of burnt rocks of mine dumps and ash-slag waste in the technology of building materials and road construction are described in [1, 6, 14, 25, 31, 34, 35, 37, 45, 65, 82, 83, 101, 123, 136, 156, 170].

2.2 Burnt Rocks of Mine Dumps

Huge reserves of burnt rocks of mine dumps and their prospects for increasing the mineral raw material base of the construction industry attract researchers and practitioners to this type of waste. In industrialized areas, where the population density, network of cities and towns, and transport communications have reached a high level, every year more and more difficulties arise associated with the search for ground natural materials for the construction of roadbed subgrade. In the structure of road construction works, the construction of the subgrade takes a special place. Such a construction requires a significant amount of natural soil, the receipt of which causes some damage to land use. A large reserve for replacing soils is dumped mine rocks.

The positive results of some researchers confirm the possibility of using burnt rocks of mine dumps in various sectors of the national economy. Burnt mine rocks are available in all coal basins. For their widespread use, technologies are needed that ensure the production of quality goods, the accumulation and generalization of practical experience, and the development of regulatory documents for their use.

Burnt rocks are promising raw materials for road construction, which could become the largest consumer of this waste. This is a justified and reliable way to use them. However, the use of burnt rocks in road construction is less common in comparison with ash-slag waste. Burnt rock can be used as a road-building material or as a raw material for its production. Practical experience in the use of coal in road construction is available in all coal regions. A significant amount of burnt rocks are used at the base of the roads. Burnt rocks compact well during rolling. In the result of rolling, part of the rocks is destroyed. The resulting fines exhibit hydraulic activity. The results are intense cementation of the entire massif and formation of a monolithic and solid base. Mine rocks are used as a soil material in the construction of subgrade hydraulic structures, site planning, reclamation of quarries, construction of dumping sites, and vertical layouts for construction [84].

A section of the Perm-Solikamsk highway was built in the Kizelovsky coal region of Russia. At the base of the road, burnt rock of the Severnaya mine was used, reinforced with cement. The mixing of the rock with cement was carried out directly at the site under construction with the help of a grader. A survey conducted after several years of operation of the road showed a good condition of the site.

The possibility and effectiveness of using a number of local materials, including burnt rocks, in road construction are substantiated in [14, 34, 37, 84, 87, 106, 118, 128, 169]. It has been established that crushed stone and sand from burnt rocks treated with binders have high deformation, strength indicators, and thermal insulation properties. Due to such materials, under freezing conditions, it is possible to reduce the thickness of pavement layers by up to 30%. Such materials can be used in the construction of roads of any categories in all climatic zones. Fine-burnt rocks are suitable as a microfiller for asphalt concrete, and mixed with lime; they are used to strengthen slopes.

Experience is being gained on the use of burnt rocks in road construction in the Rostov region of Russia. A section of the access road to the industrial site of the mine ventilation shaft is exploited in the village Kiselevo. The lower layers of the base are made of the burnt rocks of Kirov and Vorovsky mine dumps. At the mine of the 50th anniversary of the October Socialist Revolution, crushed stone mixture of burnt rock with a particle size greater than 20 up to 150 mm was distributed on an access soil road. The layer thickness was 15 cm. The access technological road is intended for the movement of mainly freight transport and transportation of mine rock. Tests of crushed stone from burnt rocks at the base of roads of category IV and V with the installation of lightweight, transitional, and lower types of coatings were carried out at the Almaznaya and Zamchalovskaya mines [25, 156]. At the experimental sites during the reconstruction and repair of industrial sites of some mines, and arrangement of technological and access roads for freight and passenger vehicles, nonmetallic materials from burnt rocks of mine dump No. 26 were used for the construction of asphalt concrete coatings on industrial sites and sidewalks of city streets. The effectiveness of the use of ore materials from mine dumps is confirmed by the positive results of the operation of experimental sections of roads, industrial sites, and sidewalks.

When constructing the subgrade of the access railway from the Sadkinskaya mine to the public railway [27, 28, 33], the rocks of the mine dumps No. 17 and Sevryugovskaya located in the village Sinegorsky, Rostov region (Russia), were used. To construct the subgrade of the access railway track, the mine rocks of the dumps were averaged by multiple transshipment with a bulldozer during shipment from the dump. The technology for the construction of subgrade from mine rocks is similar to the layered construction of the embankment from the traditional soil used.

In the world, the rocks of mine dumps are used instead of soil for engineering work (vertical planning, construction of foundations of buildings, and restoration of adjacent territories). Such rock properties as compressibility, shear, frost resistance, and corrosion resistance are close to the properties of coarse-grained natural soils.

In the USA, Germany, Great Britain, and other countries [51, 106, 108, 149, 163], mine rocks are most often used as building materials for hydraulic constructions (dams, dikes): for example, dams enclosing sludge traps for coal preparation plants in the Appalachian coal basin (Pennsylvania, Kentucky, West Virginia in the USA) and dams that prevent flooding during floods of river valleys of the Ruhr district (Lippe, Emscher, and others in Germany).

The positive experience with the use of mine rocks is confirmed by the fact that no accidents and deformations of highway and railway tracks passing through hydraulic structures were observed. To ensure a strong structure with low porosity in the material being laid, the fraction content of less than 2 mm should be at least 20–40%.

On the largest scale, coal wastes are used in Donetsk (Ukraine) [49, 122, 126]. There is experience in the construction of a bypass road using mine rocks to the village Belorechinsky. The length of the site is 500 m, the width of the carriageway is 7 m, and the traffic intensity is up to 200 cars per day. Well-burnt rock with a size of pieces 20–45 mm was used in the construction of the subgrade. The road has been successfully operated for over 20 years.

Burnt rocks are suitable for the building of subgrade without any restrictions. They have been used in the construction of the subgrade bypass road Orekhovka–Ivanovka. On the constructed section of the road, the height of the embankment is 0.8–1.5 m. The rock was poured and rolled layer by layer (25–40 cm) with smooth rollers, and no wetting was applied. During compaction, large pieces of rock were destroyed and a dense structure stable in composition was formed.

In most developed countries, mine rocks are not stored for a long time on the earth's surface. In the USA, Poland, and Germany, waste rock from processing workings is reprocessed, mixed with additives (for example, sand), and then put back into the worked-out space [3, 10, 54]. The land is reclaimed under dumps.

However, before removing the dumps, they are evaluated whether all the useful elements have been extracted from them. Extraction of aluminum, germanium, scandium, gallium, yttrium, and even zirconium from the dumps being strategic important materials can be carried out by dividing the raw materials into fractions, for example, by the electrostatic method [55].

Based on the results of existing domestic and foreign developments, it is possible to determine the purpose of burnt rocks in road construction for:

(i) Building of subgrade road and railways with conducting insulating measures.
(ii) Foundation bases for roads without treatment or with strengthening by mineral binders.
(iii) Foundation of structural layers of pavement of III–V categories in all climatic zones.
(iv) Coating auto-roads by using asphalt concrete: In asphalt concrete, burnt rocks can be used as a fractionated aggregate and a fine dispersed fraction as a mineral microfiller.

2.3 Ash-slag Waste

The problem of utilization and useful application of ash-slag waste is one of the acute problems of energy related to the economy of the industry and the environment. At 172 coal thermal power plants (TPPs) in Russia, more than 123 million tons of solid fuel are burnt per year. The annual output of ash-slag from TPPs operating on solid coal fuels averages more than 20 million tons. Accumulation of slag in dumps is about 1.5 billion tons. By 2030, the volume of accumulated ash-slag waste in Russia may exceed 2 billion tons. According to various estimates, the cost of building a new ash dump is from 5 to 18 billion rubles.

Ash dumps are a source of serious environmental and economic problems. They occupy more than 28,000 hectares. Part of the ash dumps, as the territories were urbanized, turned up in residential areas. The ash-slag dumps with a large accumulation of waste in them are able to form their own technogenic natural ecosystem, which, when interacting with the environment, has a negative impact on the natural environment. Dusting and filtering of ash dumps is a source of danger to the health of the population, flora and fauna. Dumps located near rivers and lakes are hazardous to the aquatic environment.

The world has accumulated vast experience in the use of ash and slag. Russian and foreign experience shows that road construction can be one of the most promising areas for the utilization of ash-slag waste. Roads are an essential part of the overall infrastructure. The quality of life of people in any country depends on the condition of the roads. Building of roads using ash and slag materials was carried out in various regions of Russia, especially in areas experiencing a shortage of traditional road-building materials (crushed stone, sand, cement). Production tests of experimental pavements confirm the effectiveness and feasibility of using slag mixtures in the structural layers of pavements. In Russia, more than 300 technologies for the use of slag materials in various fields have been developed. So, in the Moscow region about 300 km of highways was built using ash-slag mixtures. The ashes of the Barnaul State District Power Plant are used in structural layers of pavement on highways of categories III–IV. Ash-slag mixtures from dumps and mine rocks were used for the construction of the subgrade in the construction of access roads and local roads in many regions [27, 28, 33, 87, 128, 169]. In the Rostov region of Russia, fly ash and ash-slag mixtures [25, 27, 48, 56, 57, 156], both independently and in combination with other components, are used to strengthen small one-dimensional sands, loess soils, and low-strength limestones as a low-activity binder or particle-size additive. It is advisable to use ash-slag mixtures treated with binders in the upper part of the subgrade due to their high thermal insulation properties, which can reduce the freezing of the subgrade.

In the result of the use of water disposal slag materials, reinforced with inorganic binders, namely cement or lime, for the construction of layers of pavement, two areas of use for dump water disposal slag mixtures were determined:

(i) Treatment with cement or lime and use as structural layers of pavement
(ii) As an additive to binders in order to save them at strengthening soil

The road industry needs mineral powder for asphalt mixtures. The authors' studies [19, 63, 64, 93, 114] have showed that due to the hydraulic activity and dispersion of fly ash, it can be used as a filler in asphalt concrete with great technical and economic effect. The authors conclude that active fly ash is a good mineral powder for asphalt concrete.

The use of ash-slag waste in the near and far abroad is widespread. Among the CIS countries, the greatest experience in the disposal of ash, slag, and coal mining waste in road construction has been accumulated in Lugansk and Donetsk regions (Ukraine) [49, 122, 126, 153, 163]. The use of slag materials from the Burshtyn State District Power Plant (from burning coal of the Lviv-Volyn basin) allowed replacing up to 40% of fine-grained natural sand in concrete for the production of road slabs, curbs, paving slabs, and other products.

Ash-slag mixtures of water removal with lime additives have been used in road construction in Kazakhstan in the building of more than 300 km of roads using an ash-lime binder.

In developed countries, industrial wastes are regarded as technogenic raw materials and are exported. In most countries, waste management is defined at law level. In many Western European countries, the level of utilization of dry ash is from 70 to 100%. In the UK and Germany the entire annual output of slag is used. The leading place among the countries of Western Europe in solving the problem of the use of fuel waste from thermal power plants in road construction is held by France. Fly ash is used in all elements of road constructions, as man-made soil; as a mineral material, reinforced with a hydraulic binder, in the lower and upper layers of the base and in asphalt concrete pavements; and as a mineral powder. Benefits are provided for consumers of ash-slag waste. In India and the USA there are adopted rules on the mandatory use of ash waste in road construction.

Studies in England have shown that fly ash and ash-slag mixtures are materials suitable for the construction of embankments and the construction of the lower layers of the base of pavement. Several embankments of slag mixtures were built in Hungary. In Poland, research and experimental work has been carried out to strengthen the fly ash of sand and clay soils. In Japan, a mixture of steelmaking slag and fly ash is used for road construction. In Italy, coal ash is used as a natural filler and binder in pavement constructions. In China, in the construction of the highway, a mixture of lime and coal ash was used as a carrier layer. In Finland, fly ash is used effectively in asphalt concrete mixtures as an additive to lime fillers. With its help, marshy soils were strengthened in one of the road sections. In India, fly ash is used to strengthen the soil of the embankment and to cover. In the USA, soil reinforced with ash was used under the pavement base. In Belgium, fly ash is used as an active additive in pozzolanic concrete and as a binder component to strengthen sand.

An analysis of Russian and world experience in the utilization of waste from the coal industry and heat power industry indicates that there is a pronounced tendency to their use. Since these are already extracted promising raw materials, they are profitable to use while solving economic, environmental, and social problems. However, the amount of unused burnt mine rocks and ash-slag waste, both in Russia and abroad, is huge, despite the fact that the economic and environmental significance of their processing is obvious. These materials must be transferred to the status of "products." In this status, they can be included in the project documentation and their demand will increase.

Chapter 3
Ways of Improving the Structural Properties of Burnt Rocks of Mine Dumps and Ash-slag Waste

Abstract This chapter presents the ways to improving the structural properties of burnt rocks and ash-slags. First, technological recycling schemes of complex enrichment of rock materials with various strength are discussed taking into account different stages of technological process. In particular, the technological operations of crushing of raw materials, sorting of crushing products, and transportation and storage of finished products are considered. Equipment for the crushing and screening complex is discussed in some details. Main characteristics of crushed stone quality are studied at each technological stage. Then the results of characterization of ash-slag mixtures used in road construction are present. The obtained data are based on the practical experience of studying burnt rocks, fly ashes, and slags from dumps of several Russian mines.

Keywords Structural properties · Recycling schemes · Conditioned raw materials · Strength and durability · Screening and crushing · Mineral fillers · Asphalt concrete · Ground fraction · Ledges of mine dumps

3.1 Burnt Rocks of Mine Dumps Recycling Schemes

When dumps are formed during unloading of rock massif, large pieces of rock are usually located at the base of the dump, and its top is composed of rocks of small fractions. When studying the granulometric composition of the rock massif in the dumps, three zones are identified along the slope of the rock dump: (i) the area, located at the bottom of the dump and consisting of large pieces of rock; (ii) the region, located in the middle part of the dump, composed of medium-sized pieces of rock; and (iii) the third zone in the upper part of the dump and formed of small pieces of rock. In general, the rocks that make up the dump are heterogeneous in mineral composition, roasting degree, and structure, and have different strength and fineness: from blocks and sintered conglomerates to small pieces. Such raw materials without preliminary preparation are unsuitable for use.

A. A. Lyapin et al., *Improving Road Pavement Characteristics*, Innovation and Discovery in Russian Science and Engineering,
https://doi.org/10.1007/978-3-030-59230-1_3

Mine rock processing operations are energy intensive and expensive. To obtain conditioned raw materials from burnt rocks of mine dumps, they should be processed, the basis of which is enrichment in strength. Strength enrichment includes a complex of production operations that provide the material with the necessary strength and durability. During enrichment, separation occurs of those particles that reduce the quality of the material, and separation of the remaining material into strength classes. In this case, enrichment is carried out according to the principle of selective crushing. Strength enrichment principle is based on the separation of fine fractions from materials, the bulk of which are weak differences. Therefore, if, after crushing, the fine fraction is sieved from the products, the remaining oversize product will turn out to be more homogeneous and of more strength.

The choice of the enrichment method is determined by the quality of the raw materials arriving for processing and product requirements. The choice of methods is influenced by the working conditions of enterprises. First of all, the climatic conditions of the region and the duration of the season of the enterprise are an influence. Temporary enrichment plants should be simple in design, with low energy consumption, being transportable and uncomplicated in maintenance. In stationary enterprises, enrichment plants of a capital type are arranged. When processing rocks, one should choose such a concentration scheme that would ensure effective separation of the material by strength at minimal cost.

According to the physico-mechanical properties, burnt mine rocks can be attributed to type III rocks, which include nonuniform-in-strength low-abrasive rocks containing weak differences. Processing of such rocks causes the greatest difficulties.

Figures 3.1 and 3.2 present schematic diagrams of the technology of complex enrichment of materials with various strength. When drawing up such schemes, it is necessary to attempt to reduce the stages of crushing. Secondary crushing is provided as mandatory; the following stages of crushing are established as necessary, as a rule, in stationary units. The technological scheme (Fig. 3.1) includes the following operations:

(i) Preliminary screening that removes fines and distinguishes two classes of material size
(ii) Separate primary crushing (I) in the respective types of rotary crushers
(iii) Preliminary and calibration screening before the second stage of crushing (II), which selects a commercial fraction of crushed stone of 40–70 mm and transfers crushed stone below 40 mm into the sublattice product
(iv) Repeated screening of the under-sieve product (0–40 mm) to obtain marketable fractions of crushed stone of 10–20 and 20–40 mm
(v) The second and third (III with closed cycle) crushing stages of gravel 0–40 mm
(vi) Screening and cleaning of commercial fractions of sand and gravel

It is recommended to determine the size of the material selected for preliminary screening as follows: for the first stage of crushing it should be close to the width of the feed opening of the crushers; for the second and third stages, it should be equal to the largest particle size of the products obtained.

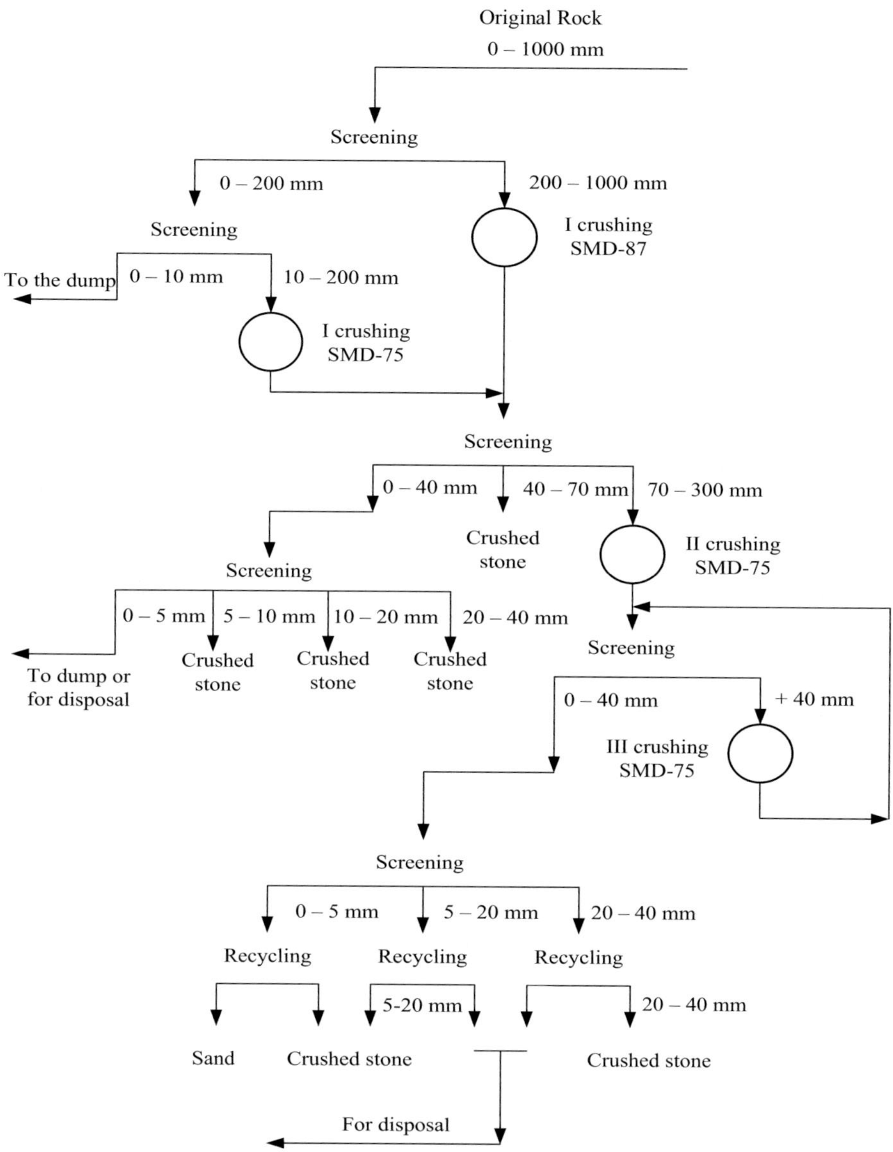

Fig. 3.1 Technological scheme of enrichment of materials with various strength

Qualitative sands can be obtained by enrichment and fractionation. The enrichment of sand consists of the removing grains larger than 5 mm; washing dusty, silty, and clay particles; and improving the grain composition.

When processing rocks containing up to 20% of weak differences, the yield of crushed stone is about half of the rock mass. Wastes obtained in large quantities in the form of screenings should be utilized as raw materials for the cement and

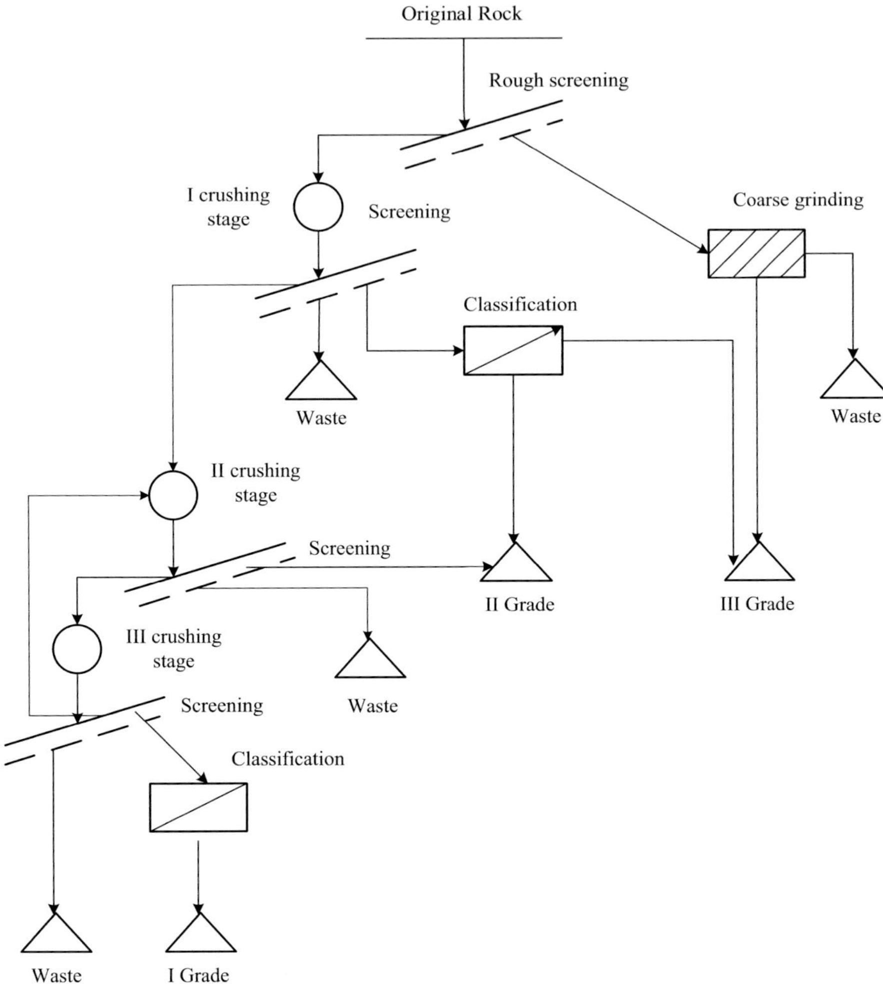

Fig. 3.2 Technological scheme of rock processing by selective enrichment by strength

ceramic industries, mineral fillers for asphalt concrete, etc. Thus, the integrated use of raw materials is achieved. If the initial rock contains more than 20% of weak differences, crushed stone of the required grade can be obtained only by including in the technological scheme a special enrichment operation for strength. If during the processing of the dump selective extraction of strong and weak rocks is possible, then for their processing it is advisable to provide separate production lines.

Figure 3.2 presents a complex scheme of processing rocks into three grades of crushed stone and waste. According to this scheme, the rock mass is sent first for rough screening. The under-sieve product is fed to a selective crushing plant or classifier to extract strong differences of the grade III product and waste. Oversize product gets into the crusher, where it is crushed and sent to a vibrating screen, where the material is divided into two fractions. The fine fraction (sublattice) is sent for enrichment to obtain crushed stone of grade II in strength and product of grade III. A large fraction (oversize) is again crushed and screened, while the under-sieve product (mainly grade II material) is sent to the warehouse, and large crushed stone is again crushed and dispersed into grades that are classified by strength. Wastes of this classification relate to products of the grade II, and the rest of the material is a product of grade I. This is the scheme for processing rocks with a three-stage crushing of rock mass.

Based on the experience of studying mine rocks available in the Rostov region (Russia), it can be assumed that the largest number of them will be used in road construction. To obtain a conditioned material for the device of the lower layers of the basics of pavement, it is only necessary to sort it by size and separation by strength classes with the allocation of low-strength differences. The technological scheme of such rock preparation is shown in Fig. 3.3.

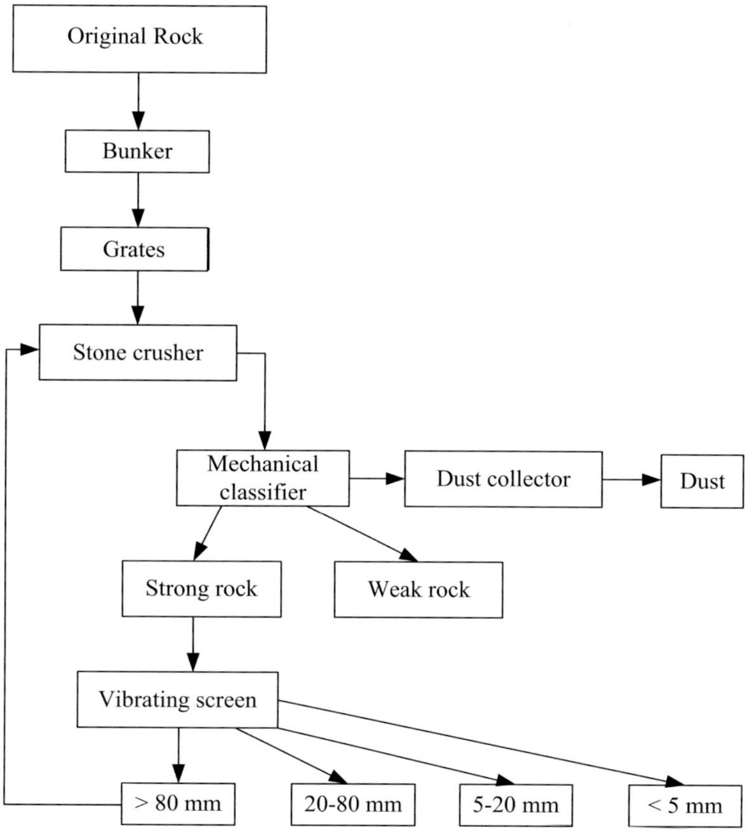

Fig. 3.3 Technological scheme of a sorting plant for processing burnt rocks of mine dumps

High-strength crushed material with a rough surface is required for the upper layers of the bases and coatings of pavements and the device of wear layers. For road concrete, a high-strength aggregate material with a predominance of cuboid shape of grains is also required.

The technology for enrichment of mine rock without crushing large fractions is as follows. The rock mass from the dump is sent to a bunker with grates, on which large material is separated. This material can be crushed according to a separate technological scheme. The main bulk of the rock after separation on the grates passes to a vibrating screen, where the sand is screened out and the material is sieved into fractions. If it is necessary to separate the material according to the strength, the fractions pass to classifiers, on which weak differences are first separated, and then they are divided into two grades according to strength.

When crushing large material, the following scheme is usually used (Fig. 3.4). The rock from the dump is fed to bunker I, where large pieces are separated, which enter the skirting board, or large crushing, into the crusher. The undersize product is conveyed by a conveyor to a vibrating screen 2 to separate sand of 0–5 mm. The rest of the mass passes to classifier 3, where low-strength differences entering the dump are separated. The enriched gravel is conveyed by a conveyor to a vibrating screen, the under-sieve product of which is sent to the warehouse. Large material from this screen is fed to the crusher for grinding, from where it is fed by a conveyor to a vibrating screen, on which sand (0–5 mm) and large grains are separated. The sand is sent to waste, and large grains to the classifier, dividing the material into three

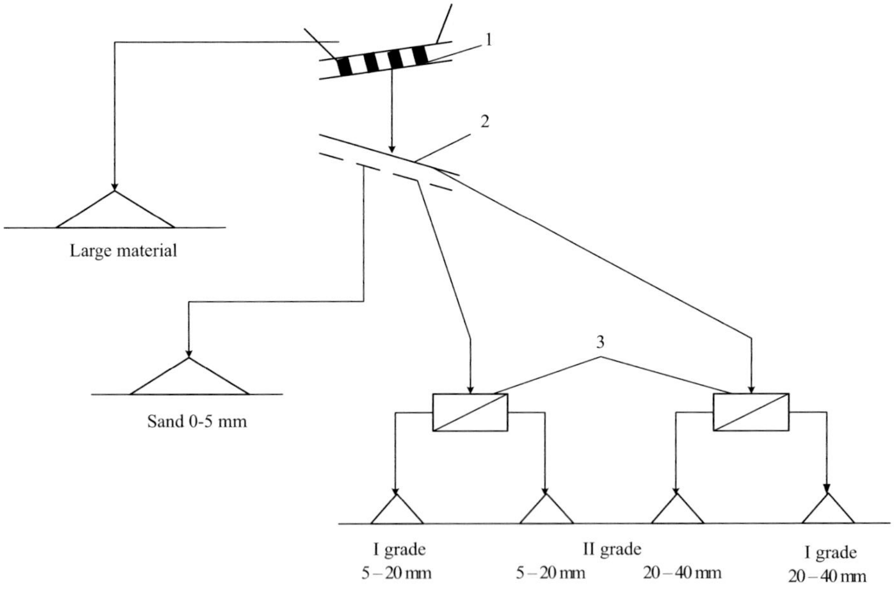

Fig. 3.4 Technological scheme of processing the burnt rocks of mine No. 26 dump

grades. When processing mine rocks, along with crushed stone, it is possible to obtain high-quality sand from crushing screenings, subjecting them to classification for separating the dust fraction of less than 0.14 mm.

This scheme is acceptable for the preparation of material for the construction of two-layer concrete and asphalt concrete pavements. It can also be used to prepare the material for the construction of lightweight improved coatings in the case when the upper layer uses more durable material, subjected to crushing and enrichment.

The above technological schemes for processing mine rocks are of a general nature. When processing a dump, individual schemes are developed for specific conditions and taking into account the properties of the rocks and the requirements for the manufactured products. For example, at the Tekhnostroy enterprise, a crushing and screening complex was mounted on the base of the rocks of mine No. 26 dump, operating according to the scheme shown in Fig. 3.5 and designed for processing of burnt clay rocks of argillite-siltstone compaction into crushed stone. The rock strength ranges from 10 to 200 MPa, and their processing is carried out according to a two-stage scheme. The main areas of application of the products are road and mine construction. Crushing screenings are used for reclamation of quarries and filling of burning dumps, and exhaust additives for brick and ceramic production.

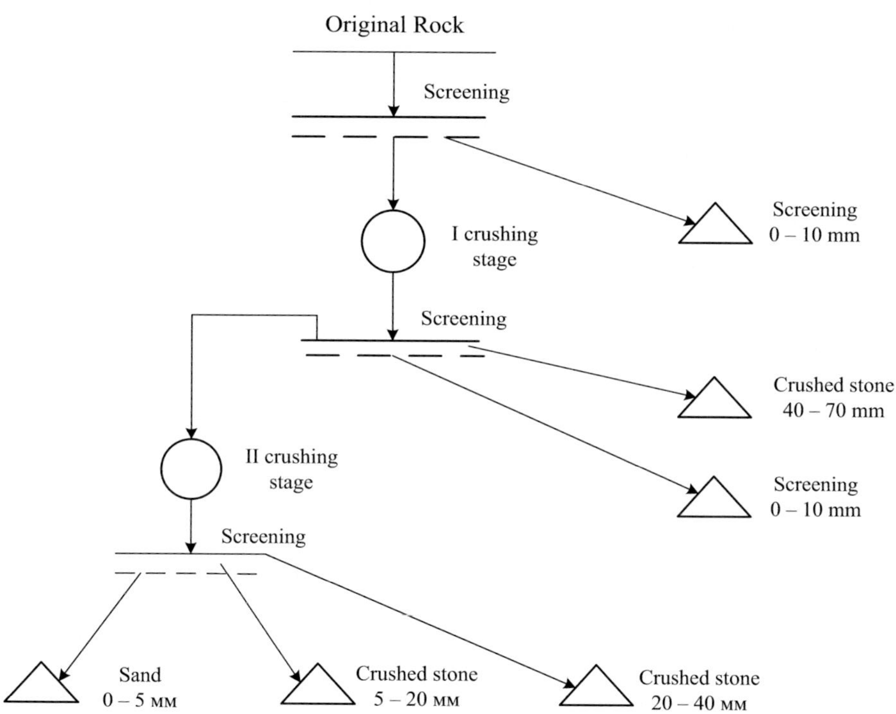

Fig. 3.5 Technological scheme of processing the burnt rocks of mine No. 26 dump

The literature provides descriptions of technological schemes for the primary preparation of burnt rocks before using them in concrete and ceramics. For example, in one of these technological schemes, a comprehensive processing of burnt rocks is carried out by selective crushing, enrichment, and disposal of waste. According to this scheme, the initial rock is first subjected to preliminary enrichment, after which the medium-sized rock is sent to the impact crusher (fine crushing), and the large rock (over 40 mm) is passed for large and medium crushing. At the stage of small and medium crushing, sorting and separation of fillers for concrete and finely ground additives are provided for binders and asphalt concrete. The second technological scheme of rock processing includes the primary stage of crushing on a self-propelled crushing unit. Primary crushing is carried out at the dump. After this crushing by a dump truck, the products are fed to the receiving bunker and then to the rotor and hammer crushers by each grated screen feeder. Each of them works in a closed cycle with a vibrating screen. The technological scheme allows one to obtain crushed stone and sand of two grades (in terms of purity). Sand grinding to obtain ground additives can be carried out together with cement in a separate ball mill of small capacity. The described technological schemes provide high-quality preparation of mine rocks for use. However, the processing technology is complicated and oversaturated with equipment and requires large material and energy costs.

Preparation of mine rocks for use in ceramics compositions is reduced to obtaining a ground fraction and a fraction of not more than 3 mm. Therefore, in technological schemes for the processing of rocks, in addition to a jaw crusher, a mill should be provided. A number of works [2, 48, 120] give options for technological schemes for preparing mine rocks for plants of various capacities, describe the grinding and grinding equipment proposed for crushing moist and dust-prone coal waste, and describe the experience of its application in a number of construction plant materials.

Processing of mine rocks should be carried out near dumps. The development of the waste dump is from top to bottom. For this, shovels, bulldozers, and dump trucks are most often used. The top layer of the rock, with a depth of at least 0.5 m, is transported to the dump. After removing the upper layer, the development of the massif of the dump begins.

The development of the dump should be carried out in horizontal layers (ledges). With horizontal platforms, the best conditions are created for the operation of excavators, automobile, railway, and conveyor vehicles. The development of a ledge is usually carried out by strips parallel to its slope (runways). The front of the work can move from one border to another in parallel or along the fan. The steps within the working area are treated simultaneously with minimal movement of excavators from the ledge to the ledge. At the same time, working platforms of normal width are preserved on all horizons. The slope angle of the working side is usually not more than 10–15°. By processing within the working area, one or several groups of ledges are distinguished, which are developed sequentially from top to bottom, expanding areas with a minimum width (W_{min}) to a maximum width (W_{max}) (Fig. 3.6).

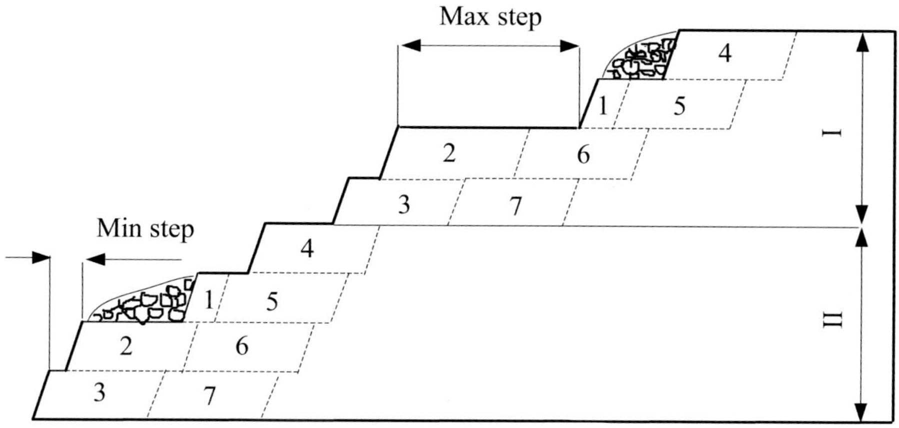

Fig. 3.6 Scheme of group processing of the ledges of mine dumps

The burnt rock is fed into the receiving bunker of the crushing plant by a bulldozer, or an excavator, or an auto dump truck. The choice of crushing and screening equipment is determined by the properties of the processed rocks such as strength, uniformity, abrasiveness, size of pieces and particles of material, and number and type of contaminating inclusions contained in them.

The main types of crushing equipment are crushers: jaw, cone, rotor, hammer, and roll. In addition to crushers, the crushing and screening complex includes screens, classifiers, conveyors, feeders, and conveyors that combine all the units into a production line. For processing mine rocks, jaw and cone crushers are applicable. However, the efficiency of selective crushing is especially high when using rotor or hammer crushers. The material obtained in such crushers, especially in rotary crushers, has a better grain shape; the yield of lamellar and needle-shaped grains is less than that in other types of crushers. Crushed stone obtained in such crushers has a higher grade in strength due to deleting weak inclusions of the original material. Rotary and hammer crushers are productive, economical in energy consumption, and easy to manufacture and maintain. Their main drawback: the need to often replace the hammers, which when crushing abrasive materials quickly fail.

Thus, to obtain marketable products from mine rocks, it is advisable to use rotary and hammer crushers for their processing. The number of crushing stages is determined by the number of aggregate fractions produced and their purpose. The largest number of fractions can be obtained with a three-stage crushing scheme (see Fig. 3.7). The above scheme uses rotary crushers of large, medium, and small crushing. In the last crushing stage, a hammer mill can be used. This scheme provides for the possibility of obtaining in large quantities a large fraction of crushed stone for roads with fractions of 40–80 (150) mm. Such a fraction can be selected after the first stage of crushing. To increase the productivity of the crusher of medium crushing, it is necessary to remove fines (0–10, 0–20 mm) from the intermediate product after the first crushing stage. If there is no great need for a large fraction, then the

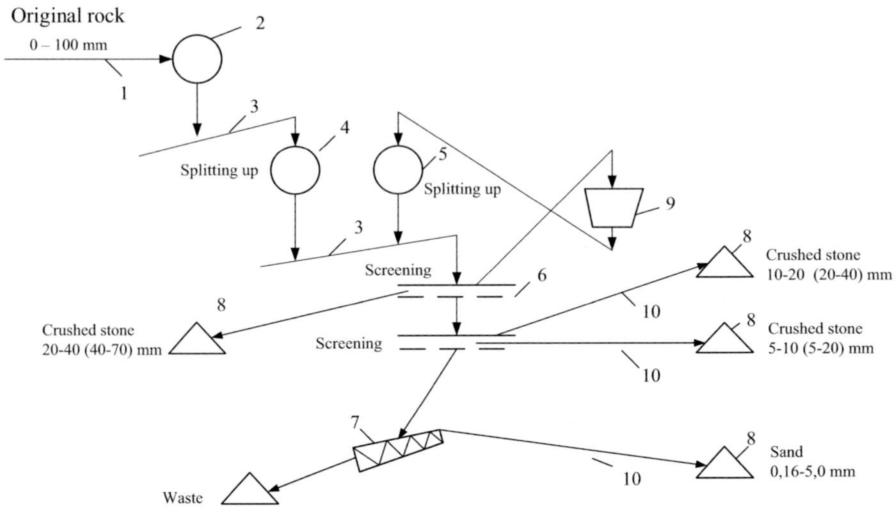

Fig. 3.7 Technological scheme of a three-stage crushing of mine dumps of nonuniform-in-strength burnt rocks: 1—bunker with a plate feeder; 2—aggregate of large crushing; 3—tape conveyors; 4—medium crushing unit; 5—fine crushing unit; 6—sorting unit with two inertial screens; 7—flushing unit with a special classifier; 8—storage bunkers with weight batchers; 9—intermediate bunker-warehouse; 10—stacker conveyors

semi-product after the first stage of crushing is sent to the next stage of crushing. After the first stage of crushing, the processing of rocks can be carried out according to one of the options:

(i) Two medium and fine crushers with a closed cycle at the last stage are included in the scheme; according to this option, three fractions of crushed stone and sand are obtained.
(ii) One crusher of medium or small crushing is included in the scheme, depending on the size of the commercial products, and crusher products are sorted on one unit; the presence of three stages of crushing makes it easy to control the output of fractions of crushed stone depending on the need and to carry out preventive work without stopping the complex.

It is economically feasible to use mobile crushing and screening plants manufactured by machine-building plants for processing mine dump rocks. Such plants consist of aggregates, each of which performs only one technological operation. It allows one to use them in various combinations for processing igneous, sedimentary rocks and gravel-sand mixtures. Depending on the material being processed, various layouts of the aggregate scheme are possible. For example, the medium-capacity plant consists of two aggregates for coarse and fine crushing and is designed to produce fractions of crushed stone of 0–5, 5–10, and 10–25 mm. With appropriate adjustment of the slots of the crushers and replacing the sieves, crushed stone with a size of up to 40 mm can be obtained. Such installations operate in Gukovo city at the dump of mine No. 26 and in Shakhty city at the dump of mine named after Petrovsky in the Rostov region (Russia).

Fig. 3.8 Crushing and screening plant for processing burnt rocks at the dump of the Mayskaya mine

The preparation of crushed stone with a grain size of up to 40 mm from abrasive rocks is provided for by a three-stage closed-cycle crushing scheme with a screen at the second and third stages. According to this technological scheme, crushing and screening plants with a capacity of 85 m^3/h work. The products of all crushers are sorted on one unit; commodity screening without washing the undersize product of the first sorting unit is carried out on the second sorting unit. Material crushing is carried out on jaw and cone crushers. A similar installation (see Fig. 3.8) is mounted and works on the dump of the Mayskaya mine (Rostov Region, Russia).

For processing sedimentary low-abrasive rocks, there are plants that include two aggregates with rotor crushers for large and medium crushing. Productivity of the installation according to this scheme is 70–100 m^3/h. The advantage of using crushing and screening plants is that such complexes are equipped with a complete set of necessary equipment for supplying, crushing, classifying, and storing finished products. Installations are equipped with an individual power plant. Dust extraction, hermetic suction shelters, and air cleaning in dust collectors are provided before it is discharged into the atmosphere.

Taking into account the difficult economic working conditions of building material enterprises, the physical and moral aging of equipment, as well as the changing requirements for the quality of crushed product, machine-building plants develop and propose new equipment [81] that meets modern production conditions. The main goal of the development is to simplify the design, reduce metal consumption and the number of quickly wearing parts, and increase the ease of installation and operation, and the efficiency of crushing, too. Modernization affects almost all types of crushers.

Most crushers are produced by factories, both in the form of independent equipment and as part of aggregates and plants, equipped with auxiliary devices and equipment. Crushing aggregates are produced for the processing of wet and sticky materials with simultaneous drying. Units are equipped with crushers, heating system, cyclones, and lock gates. There are developments of crushing and screening plants for producing crushed stone of a cuboid shape (capacity 20–40 m^3/h), and for crushing sedimentary rocks with equipment for washing and enriching crushed stone by strength (productivity 60–70 m^3/h). New crushing equipment should ensure efficient processing of rocks, reduce the cost of finished products, and improve its consumer properties.

Ash-slag waste is a complex polymineral system, the quality and properties of which are influenced by many factors: the composition of the mineral part of the fuel, the type of boiler units and the mode of burning coal, the method of ash and slag selection, etc. As Russian and world experience shows, the use of ash-slag waste is associated with certain difficulties. In the construction industry, dry ash, sand, and crushed stone from slag and an ash-slag mixture of dumps are used. It is not possible to consider waste heat power engineering as a conditional raw material directly suitable for producing high-quality products and a wide range of goods. Therefore, a preliminary classification is required.

Burning fuel at temperatures at which 90–100% of the mineral part is converted into a melt allows one to obtain a product of increased uniformity. As the result of rapid cooling of the melt in water, granules of granular slag 10–15 mm in size are formed. In such a fraction, there is an increase in the amount of glass phase in comparison with an identical ash, reaching 90–92%. Then a product is formed with increased hydraulic activity due to the glass phase. If necessary, the slag is crushed to the required size. The burning of solid fuels with the conversion of their mineral part to slag is a progressive area in the energy sector. For example, granular slag of fire-liquid removal, obtained from a variety of coals and at different temperatures of fuel combustion, has basically uniform properties. Such a product is an effective and high-quality material for the production of many building products and structures, and for road construction. Liquid slags have stable properties and are quite suitable for road concrete.

Dry ash is more uniform in composition and properties than ash from dumps. At power plants, fly ash is collected by various ash collectors, accumulated in a receiving bunker and sent to a silo warehouse, from where it is shipped centrally to the consumer in cement trucks or hopper cars.

When examining ash dumps, it was found that in depth and area of the dumps, the waste has significant heterogeneity in moisture, grain composition, bulk density, amount of unburnt carbon particles, etc. The spread in the distribution of ash and slag of various sizes ranges from 20 to 80%. Dump ash is characterized by heterogeneity of the chemical and mineral composition. This is due to the combined influence of the type of fuel burned; the condition for its preparation, atomization, and combustion; and the variable load on the boiler units. In the process of movement of the pulp, a fractionation of ash particles occurs by size, mineral composition, and particle shape. So, unburnt coal particles having the lowest density are concentrated mainly in large fractions of ash, which contain an increased amount of calcined clay material and are transported by a pulp stream over large distances. The glass phase, as a rule, is concentrated mainly in the smallest ash fractions.

A characteristic feature of dump ash is variable humidity (5–40%), as a result of which it dusts in the summer and freezes in the winter. Fluctuations in ash moisture bring production complications. They are inevitable when using ash in road construction, and in other types of production. At a certain humidity, ash is prone to clumping. As a result, it will stick to vehicles. The use of vibration to increase the fluidity of the ash does not lead to the desired results. To the greatest extent, this is manifested at a moisture content of 30–40%. If humidity is reduced to 15% or lower, then dusting during loading and unloading operations and transportation is inevitable, which essentially eliminates the possibility of using ash. The use of slag as a raw material in the construction industry under such conditions makes it difficult to produce high-quality products and promotes the formation of a negative attitude among manufacturers on the widespread use of slag. To engage in the production of large stocks of slag stored in dumps, it is necessary to presort this raw material in ash dumps. Preparation of ash-slag waste for use is less time consuming and costly than mine rocks. Ash-slag mixtures on dumps, as a rule, are developed by consumers and sent by their own transport. At the same time, selective mining of the slag mixture with the maximum slag content and its further enrichment to the optimum composition in most cases are not performed. These works should be carried out at power plants and include measures to bring waste into marketable condition that meets the requirements of state standards and technical conditions.

The development of ash dumps should be carried out in a career way. Stripping operations are carried out with bulldozers, scrapers, or excavators to remove clay, sand, or plant soil, which are laid to the surface to prevent dusting of ash dumps during conservation. To develop slag dumps, we can use a variety of excavators (single and multi-bucket, dragline excavators), as well as other machines and vehicles. At ash dumps, primary processing of raw materials is carried out, which consists of partial dehydration of ash-slag waste and bringing moisture to optimal. To do this, the ash-slag mixture is poured into a cone and only after it has matured it is sent to the consumer.

The simplest and quite effective method of averaging hydraulic dump ashes is their multiple transshipment. If it is carried out during loading and unloading operations during the transportation of ash, then additional costs for this operation are not required. In this sequence, the uniformity of the particle size distribution of the ashes increases, and their dusting during transportation and freezing in the winter are prevented.

The greatest degree of averaging of slag mixtures is achieved during enrichment. Enrichment methods are the sieving and flotation. The enrichment of slag mixtures consists of the removal of large (gravel) grains and the improvement of grain composition. Separation of gravel fractions is performed by screening on vibrating flat or drum screens or classifiers of various designs. The main purpose of enrichment is to provide the required grain composition of ash-slag waste. If necessary, after enrichment, slag is dehydrated.

The authors of [41, 59, 111, 124] describe the available theoretical and practical experience in processing ash for various industries. The authors propose controlling the properties of ash-slag waste through the use of various separation methods, chemical reactions, and solid fuel combustion modes. A directed change in the physico-mechanical properties of polymineral raw materials can be achieved by phase transformations in systems. Such transformations occur upon influence to a system of chemicals and high-temperature treatment, accompanied by a change in the chemical state of the phases and their physical and mechanical properties. As a result of directed influence on the system, conditions are created for separation of waste-free separation of raw materials and obtaining high-quality products.

In the enrichment of slag mixtures, four groups of methods for separating polymineral systems are applicable: gravitational, magnetic, electrostatic, and flotation, as well as special, taking into account the peculiarities of the raw materials and the purpose of the products obtained.

A promising direction contributing to an increase in the volume of heat energy waste utilization is the fractionation of ash-slag waste, that is, their separation by grain size into two fractions (coarse and fine), and supplying the consumer separately large slag (1.25–5; 0.63–5 mm) and small (0–1.25; 0–0.63 mm). At the ash dump, the yield of these fractions can vary significantly. However, there are two problems. The first is the choice of fractionation technology, and the second is to ensure conditions for the efficient use of ash-slag waste, divided into two fractions. It is possible to supply fractionated slag in the form of a mixture of fractions in predetermined proportions providing the required grain composition. Slag loading should be arranged at ash dumps to be shipped to the consumer by railroad or waterway. To ensure uniformity and stability of properties of ash-slag waste, it is necessary to design the most advanced methods of burning coal and select ash-slag during the design and construction of new power plants.

At the existing power plants, the ash-slag mixture of water removal again coming to dumps can be classified. To do this, it is necessary to install at the end of the pipeline a classifier with a set of screens with different sizes of holes. Water containing the pulverulent fraction is discharged into the storage. Thus, fractions are obtained in which there are no dust particles.

At foreign power plants, coal is burnt, as a rule, in a fluidized bed. Separately combined methods of ash removal are provided: the slag is removed hydraulically, and the ash is deleted pneumatically. This scheme is justified. In Russia, the pneumatic method is used far from all power plants. The low prevalence of this method is mainly due to the still insignificant volumes of utilization of ashes in the national economy. Over time, the pneumatic method should be widely used. All newly designed Russian coal-fired TPPs provide for pneumatic ash extraction and feeding of dry ash to the silo warehouse or pumping of it to ash dumps.

At a set of TPPs, the collection and delivery of dry ash directly from bunkers, located under specially raised electrostatic precipitators, are provided. Ash enters into bunker by gravity along the estrus (pipes) from the electrostatic precipitators. In designs of thermal power plants, two different systems for the selection and delivery of slag are used: with alluvium into dumps or section sludge traps. For the issuance of ash-slag mixture from the dumps, multistage sectioning is used with the following cycles being performed alternately: filling, dehydration, development by bulldozers, and scrapers with loading the mixture into vehicles with excavators.

In the Rostov region (Russia), at Novocherkasskaya state district electric station, a dry ash sampling plant is operating. The plant productivity is 134,000 tons of ash per year. Shipment to the consumer of dry ash is performed in hopper cars and cement trucks. At ash dumps, an ordinary ash-slag mixture is shipped for transportation by rail or road. A plant for the processing of ash-slag waste and the delivery of a conditioned product to consumers is being built at the ash dump.

3.2 Non-ore Building Materials from the Burnt Rocks of Mine Dumps and Ash-slag Waste

3.2.1 Properties of Burnt Aggregates from Mine Dumps

The processing of rocks into aggregates is carried out at a crushing and screening device according to the technological scheme, which includes the main technological stages of processing: (i) crushing of raw materials to obtain crushed stone fractions and a given grain size screening; (ii) sorting crushing products into specified fractions; and (iii) transportation and storage of finished products. The main equipment of the crushing and screening complex is crushers and screens, as well as tape conveyors, feeders, and conveyors that combine all the units into a production line. Figure 3.9 presents a flowchart of a crushing and screening complex operating on a waste dump No. 4 of the Glubokaya mine owned by the Donstroyresurs enterprise (Shakhty city, Rostov Region, Russia). The burnt rock with a single-bucket excavator is loaded into dump trucks and delivered to the processing site, located at a distance of 1 km from the dump. The rock massif from the dump truck is fed into the receiving hopper. From the hopper with a plate feeder, the rock is fed to a jaw crusher. The primary crushing rock enters with a maximum piece size of 100–300 mm. The maximum size of pieces of rock in the second stage of crushing is 20–80 mm.

After the jaw crusher, crushed rock is conveyed by conveyor No. 1 to an inertial screen equipped with two screens, where calibration (control) screening is performed to isolate a product from the material stream that does not require a second crushing stage. At this stage of screening, the required crushed stone fraction (from 5 to 20 or from >20 to 40 mm) and screenings are selected from crushing products in accordance with needs. After the screening, the crushed stone fraction is conveyed by conveyor No. 3, and screenings by conveyor No. 2 are stored near the crushing and screening device.

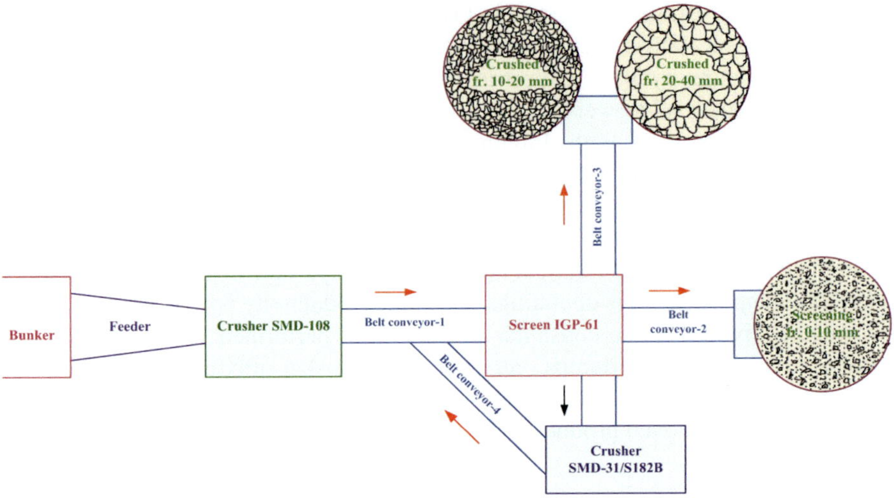

Fig. 3.9 Technological layout of the equipment for the crushing and screening complex on the basis of the mine shaft dump No. 4

Oversize product after screening is sent to secondary crushing, which is carried out on a hammer mill. The crushed product from the hammer mill through conveyor No. 4 is returned to conveyor No. 1 and sent to an inertial screen. According to this technological scheme, the products of both crushers are sorted on one screen. After the screen, the resulting product is conveyed by conveyor Nos. 2 and 3 to the finished goods warehouse. Warehousing and storage of crushed stone are carried out in an open area with a hard coating on fractions in cones. Screenings, intended primarily for use in the manufacture of ceramic bricks, are stored in a cone under a canopy. Shipment of finished products is carried out by a forklift.

The number of crushing stages is determined by the number of aggregate fractions produced and their purpose. Long-term research and production tests have established that at least two stages of crushing rocks should be provided during the processing of mine rocks in the production line.

The primary products obtained during the processing of rocks are non-ore building materials: gravel, crushed stone-sand mixtures, screenings of crushing rocks, and sand from screenings of crushing. Tables 3.1 and 3.2 present the main indicators of the physico-mechanical properties of aggregates from burnt mine rocks.

Burnt rock mine fillers (see Fig. 3.10) are not lower in quality to similar products from traditionally used raw materials, and even have some advantages. They do not contain silt and clay particles and other contaminants (clay is only in lumps). Crushed stone withstands the tests for structural stability against all types of decay. The surface of the particles is not rolled, torn, and rough, and it does not contain clay and other clogging impurities. This feature of such fillers increases the adhesion of gravel particles in the structural layers of pavements and with cementitious materials in concrete and asphalt concrete mixtures. As a result of

Table 3.1 Quality indicators of crushed stone from burnt rocks of mine dumps

Main characteristics of crushed stone quality	Value of indicators for fractions, mm			
	5–20	>20–40	>20–40	>40–80(70)
Strength grade when testing for crushability	800; 1000; 1200	600; 800; 1000; 1200	600; 800; 1000;	600; 8000; 1000
Frost resistance grade, at least	F15; F25; F50	F25; F50; F100	F25; F50; F100	F25; F50
Water resistance grade	W1; W2	W1; W2	W1; W2	W1; W2
Grade on abrasion	A1; A2	A1; A2	A1; A2	A1; A2
Bulk density, kg/m^3	1120–1240	1100–1260	1140–1290	1160–1300
True density, g/cm^3	2.66–2.68	2.67–2.69	2.66–2.69	2.65–2.71
Dust content and clay particles, %	0.90–1.50	0.70–1.15	0.08–0.93	0.05–0.08
Clay content in lumps, %	Practically absent			
The content of lamellar and needlelike grains, %	20.0–27.0	18.0–30.0	1.25–3.85	3.2–11.8
Grain content of weak rocks, %	4.50–8.0	7.0–9.0	1.77–2.30	1.1–2.7
Water absorption, %	3.3–4.5	3.0–5.5	2.70–3.86	0.05–1.00
Loss on ignition, %	3.5–4.7	2.85–3.70	1.70–1.90	2.0–3.2
Structural stability against all types of decay, mass loss during decay, %	0.13–0.75	0.25–1.05	0.32–1.27	0.65–1.33
Content of calcium oxide, %	1.15–1.40	1.13–1.24	1.07–1.18	1.03–1.36
Content of magnesium oxide, %	0.77–1.20	1.13–1.30	1.03–1.27	1.05–2.52
Sulfate content in terms of SO_3, %	1.28–2.15	1.25–1.48	1.16–1.27	1.13–1.48
Pyrite content in terms of SO_3, %	0.17–0.38	0.03–0.04	0.02–0.08	0.04–0.08
Content of alkaline oxides ($Na_2O + K_2O$), %	3.50–4.25	2.18–3.51	2.70–3.27	2.95–3.30

thermal treatment, the rock particles acquire a porous structure. However, low water absorption values indicate that the majority of pores in gravel particles are closed. The presence of porosity explains the reduced bulk density in comparison with natural gravel. Crushed stone does not contain impurities and components attributable to harmful impurities. Metamorphosed unburnt fuel particles may be present. In composition, they are different from the initial fuel and consist of coking products (semicoke and coke residues). They are resistant to oxidation and are durable when exposed to moisture and temperature changes. The gravel contains particles of a cuboid shape, as well as lamellar and needle shape. However, the content of the latter does not exceed regulatory requirements. Screenings of crushing of burnt rocks (see Fig. 3.11) according to the grain composition correspond to a sand-gravel mixture. The sand fraction of screenings for crushing rocks by physical and mechanical properties can correspond to sands of class I or II. They can be used as building sands. Modulo coarseness, sands are large and medium. Screenings are not subject to frost heaving, practically do not soak, and do not swell in water. By filtration ability, screenings are permeable.

Table 3.2 Quality indicators of screening from burnt rocks of mine dumps

Main characteristics of quality of screening (fraction 0–5.0 mm)	Values of characteristics
Grade of crushed stone fraction in strength during crushing test: Fraction 5–10 mm	400; 800
Plasticity grade: Fraction less than 0.63 mm	Pl2
Bulk density, kg/m^3	1150–1320
True density, g/cm^3	2.67–2.69
Content of dust and clay particles, %	1.58–3.75
Stability of the structure of crushed stone fraction against all types of decay, mass loss during decay, %: Fraction 5–10 mm	0.70–1.3
Filtration coefficient, m/s	6.80–7.32
Content of crushed stone fraction 5–10 mm, %	25.3–28.7
Loss on ignition, %	5.0–6.0
Content of calcium oxide, %	1.15–1.40
Content of magnesium oxide, %	0.77–1.20
Content of sulfur, sulfates in terms of SO_3, %	1.28–2.15
Pyrite content in terms of SO_3, %	0.17–0.38
Content of alkaline oxides in terms of Na_2O and K_2O, %	3.50–4.25

Fig. 3.10 Fractionated crushed stone from burnt rocks of mine dump No. 26

Fig. 3.11 Screenings of crushing burnt rocks of mine dump No. 26

According to the technical characteristics and quality indicators, gravel, crushed stone-sand mixtures, and sand from burnt rocks meet the requirements that are set by state standards for natural stone aggregates, as well as technical specifications for burnt aggregates of mine dumps for road construction. The latter sets additional requirements for fillers from burnt rocks of mine dumps. By the value of the specific effective activity of natural radionuclides (A_{eff}), crushed stone and screenings of crushing rocks belong to the first class and there are no restrictions on their use.

The widespread use of non-ore building materials from burnt rocks is constrained by the unconventionality of the material and the psychological unpreparedness of the consumer. However, practical experience in the use of crushed stone mixtures and screenings of crushing from burnt rocks in road construction and in the construction of industrial sites, as well as the improvement of territories, is already available [22, 25, 156]. Gravel, crushed stone-sand mixtures, and screenings from burnt rocks gradually conquer the market of fillers.

3.2.2 Properties of Ash-slag Aggregates

In road construction from ash-slag waste, fly ash (ash) and slag mixture are used. Dry ash is a fairly homogeneous material in terms of its chemical and granulometric composition and dispersion and has a certain activity. Fly ash is used for the construction of reinforced road bases and coatings as

(i) An active hydraulic additive in a mixed binder in combination with cement or lime
(ii) Independent slowly hardening binder
(iii) Mineral microfiller in the composition of asphalt concrete

The main requirements for ash are the following:

(i) Humidity, not more than 1.0%.
(ii) Specific surface, not less than 150 m^2/kg.
(iii) CaO content, not less than 10.0%.
(iv) MgO content, not less than 5.0%.
(v) SO_3 content, not less than 3.0%.
(vi) $Na_2O + K_2O$ content, not less than 3.0%.
(vii) Loss on ignition within 10.0–20.0%.
(viii) The specific effective activity of natural radionuclides must comply with the requirements of regulatory documents that establish the boundaries of their application.

A distinctive feature of slag mixtures of water removal is the heterogeneity in grain composition. The slag mixture from dumps contains fillers from coarse and fine grains (slag crushed stone and slag sand), as well as pulverized ash. Table 3.3 shows the results of particle size distribution of slag mixtures of samples taken from different points of the hydraulic dump.

During water removal, larger and heavier particles of slag settle near the nozzle of the slurry pipeline, and smaller particles of ash settle on the periphery. As a result, on the dump there is a large heterogeneity of ash and slag waste according to particle size distribution.

By grain composition, crushed stone and sand from slag are divided into ordinary (unsorted mixture of slag crushed stone and sand) with a grain size limit of 20 mm, fractioned crushed stone with a grain size of 5(3)–10 mm and 10–20 mm, and slag sand with a maximum size grain of 5(3) mm. In the crushed stone fraction over 5 mm, dry ash in the slag mixture may contain from 15 to 30% and the size modulus of the sand fraction can vary from 1.5 to 3.7. The requirements for the

Table 3.3 Grain composition of ash-slag mixture of averaged probe

The size of lattice cells, mm	Sieve residues, %	
	Partial	Full
15	2.25	2.25
10	5.5	7.75
5	13.5	21.25
2.5	23.5	47.75
1.25	5.5	50.25
0.63	7.75	58.0
0.315	5.5	63.5
0.14	13.0	76.5
< 0.14	23.5	100

Table 3.4 Characterization of ash-slag mixture from dumps of Novocherkasskaya SDPS

Main characteristics of quality	Values of characteristics	
Specific gravity, g/cm^3	2.45	
Volumetric weight, kg/m^3	1390	
Volumetric weight of the skeleton, g/cm^3	1.72	
Grade of crushed stone fraction (in dry state) in strength at crushability test		
From >5 to 10 mm	300	
Water resistance grade	W1	
Grade of crushed stone fraction on frost resistance		
From >5 to 10 mm	не менее, F25	
Sieve analysis, residues on sieves, % by weight	Partial	Total
10 mm	11.52	11.52
5 mm	13.43	24.95
2.5 mm	15.14	40.09
1.25 mm	10.13	50.22
0.63 mm	12.27	62.49
0.315 mm	8.62	71.11
0.16 mm	13.65	84.76
<0.16 mm	15.24	–
Resistance of crushed stone fraction against all types of decay, mass loss during decay, %	0.78	
Humidity at the yield strength, %	26.8	
Humidity at the rolling boundary, %	25.1	
Plasticity	1.7	
Consistency (consistency indicator, W)	W = 0.24	
Group by degree of bulging	I (non-bulged)	
Filtration coefficient, m/day	3.75	
Porosity	41.0	
Compaction ratio	0.98	
Density of dry soil with standard compaction, g/cm^3	2.13	
Optimum humidity with standard compaction, %	18.8	
Angle of internal friction, degrees	24°54′	
Adhesion, MPa	0.008	

content of oxides are the same as for fly ash. Slag crushed stone should be resistant to all types of decay. Grade for frost resistance should be not less than F15 and for strength not less than 20 MPa.

Table 3.4 shows the quality indicators of ash-slag mixtures stored in ash dumps of Novocherkasskaya state district power station (SDPS) of the Rostov region (Russia).

The ash-slag mixture, the value of the relative deformation of frost bulging which does not exceed 0.035, and the value of the filtration coefficient of at least 0.2 m/day can be used to arrange frost protection and heat-insulating layers of pavement.

Chapter 4
Composition and Properties of the Burnt Rocks of Mine Dumps and Ash-slag Waste

Abstract This chapter treats composition and properties of the burnt rocks of mine dumps and ash-slag waste. Beginning with formation sampling, physical and chemical properties and granulometric and mineralogical compositions of samples from different Russian mines are studied by physico-mechanical methods. Then hydraulic activity of ash and burnt rocks of mine dumps, allowing one to use these wastes as active mineral additives, is studied. Kinetics of the involvement of additives in the reaction with calcium hydroxide are discussed. Changes in the specific surface of the additive composition and in the adsorption activity of a mixture of fly ash and burnt rocks are considered.

Keywords Testing the rocks of mine dumps · Quartering method · Physico-chemical and mechanical analysis · Perlitic and siliceous natural cement · Granulometric composition · Mineralogical composition · Diffraction pattern · Thermogram

4.1 Sampling and Research Methods

The greatest difficulty is the testing of mine rocks stored in dumps. It is technically difficult to assess with an adequate degree of reliability the quality of stored rocks in the entire dump volume. In practice, only the surface layer of the dump is studied (to a depth of 4 m). Therefore, unlike the testing of current waste, the data obtained for dumps cannot be considered representative enough. These shortcomings are inherent in both Russian and world methods of testing the rocks of mine dumps. For example, in the UK, using drilling equipment, samples are taken from the surface of the dump to a depth of 3 m.

Testing of rocks of mine dumps and ash dumps is carried out according to the methods of [95] and [103, 110], respectively. The dump is divided into zones from which a primary sample is taken, composed of portioned samples. Depending on the

A. A. Lyapin et al., *Improving Road Pavement Characteristics*, Innovation and Discovery in Russian Science and Engineering,
https://doi.org/10.1007/978-3-030-59230-1_4

surface area of the zone, the number of samples varies from 3 to 5. The separation of dumps is divided into zones from its top to the base. The vertical distance between the zones is 4 m for a conical dump and 1 m for a plane dump. The lower zone of block (more than 1.0 m) and large-boulder (0.5–1.0 m) material is not subject to testing. Point sampling from the surface of mine and slag dumps in a handful way or by scooping. The method of testing the dump by the point method differs from others in the simplicity of the selection of waste material and the quickness of the implementation of the testing work.

Sampling was carried out uniformly tiered throughout the entire mass of the dump. The sampling depth was not less than 0.5 m, and the mean distance between samples was 25–30 m. Averaged samples were composed of point samples by thorough mixing. At each dump, at least 35 group samples are formed from the selected material by the quartering method. The total mass of the sample was at least 300 kg. Minimum mass of laboratory test material: for chemical analysis it was 100–200 g and for technological tests it was 50–200 kg. To study the additional characteristics of the rock (the behavior of crushed stone in road concrete, the determination of wear in the drum), gross samples in the form of crushed stone with a mass of 25–50 kg are selected. To determine the gross output of salable stone, gross samples of 50–75 m^3 are taken. Samples for laboratory and analytical studies were taken from the averaged sample by the quartering method. For physicochemical, physico-mechanical, and technological tests, averaged samples were prepared in accordance with the requirements for the preparation of samples for certain tests. For research, we used a set of methods of physicochemical analysis, including mineralogical-petrographic microscopy, light microscopy, X-ray diffraction and differential thermal analyzers, as well as physico-mechanical methods. A comprehensive analytical and technological assessment of technogenic raw materials was carried out with the aim of completeness and complexity of its use and environmental safety. Rational combinations of research methods allow us to develop optimal technological schemes for processing burnt rocks to eliminate those signs that impede the use of this raw material in the selected direction, improve its quality, and determine industrial value. It is possible to establish the effectiveness and feasibility of using products from burnt rocks of mine dumps to ensure the bearing capacity and reliability of the foundations and structural layers of pavement.

4.2 Burnt Rocks of Mine Dumps

4.2.1 Formation of the Burnt Rocks of Mine Dumps

Burnt rocks are a product of long-term self-firing of primary mines. The initial materials for their formation are metamorphosed carbonaceous rocks of carbon age, which were formed as a result of dehydration and cementation of primary clay minerals. If combustion occurs in the bowels of the earth as a result of spontaneous combustion of coal occurring in inter-coal clay rocks, natural burnt rocks or "slabs" are formed. Such burnt rocks at the exits of coal seams fall into the overburden, but usually require self-development, being a mineral, accompanying solid fossil fuels.

In coal mining, intergonal rocks give out to the surface of the earth, where they form huge cone-shaped or plane dumps. During oxidative self-firing of mine rocks, extracted together with coal to the surface, burnt rocks are formed in dumps. Since the rocks of various mine strata are usually sent to dumps, and sometimes coal treatment waste, the composition and properties of combustion products vary over a wider range than that of natural burnt rocks (slabs).

The reason for the ignition of rock dumps is the presence in the dump mass of combustible substances: pure coal, intergrowths of coal with the rock, wood, and sulfur. The propensity for spontaneous combustion of coal is different and depends on the degree of their metamorphism, humidity and ambient temperature, and duration of the period of storage of the mass in the dump. The highest reactivity is possessed by high-ash coals and carbonaceous rocks. Their carbonaceous substance is represented mainly by fusion, which is characterized by easy inflammability. Pyrite has a significant effect on the oxidation rate. Its role is to activate the carbonaceous substance as a result of its sulfonation with sulfuric acid, which is formed during the oxidation of pyrite. The likelihood of spontaneous combustion and intense combustion is higher if the rocks are large intergrowths and slowly erode, which creates conditions for oxygen to enter the deep dump zones. According to available data [89], the combustion process in dumps occurs at temperatures from 600 to 1500 °C and lasts for decades. This creates the conditions for complete firing of the entire dump mass.

In burnt and burnt-out dumps, the rock mass has a different degree of thermal variability. This is due to the peculiarities of the material composition of the rocks, irregular nature of the distribution of local focus of burnout, and different burning intensities. Based on this, burnt, burnt-out, remelted rocks and black blocks are distinguished in burnt dumps [43]. Actually, burnt rocks should be called completely burnt empty mine rocks containing a minimum (less than 5%) amount of carbonaceous impurities and a mineral, usually clay-sandy, calcined part to some extent. Depending on the degree of firing (from complete sintering to barely noticeable calcination) and the mineralogical composition, the burnt rocks have different colors: black, brick red, pale pink, and yellow. Dark and grayish-brown color indicates that the burnt rock is poorly burnt and retained part of the coal. A pinkish, yellowish, or varying saturation brick color indicates that the rock is well burnt and does not contain carbonaceous inclusions. The presence of iron and titanium oxides greatly influences the color. Iron oxides impart a brick-red color to the rock, and titanium oxides change the red hue from yellow to red-brown. At a firing temperature of about 1000 °C, a process of reduction of Fe_2O_3 to FeO and even metallic iron can occur, giving the rock a metallic gray color.

4.2.2 Physical Properties of the Burnt Rocks of Mine Dumps

This section presents the results of a survey and study of the composition and properties of burnt rocks of some of the largest mine dumps available in the Rostov region (Russia). The study of burnt rocks is carried out in two stages:

(i) Preliminary (search and revision), revealing objects for further study as raw materials for various sectors of the economy
(ii) Final (technological, industrial), as a result of which initial data are prepared for designing a device or enterprise for waste processing

The systematic approach [102], being a necessary methodological tool, is the basis for assessing the quality of these technogenic raw materials. This approach consists of obtaining primary information on the chemical and material composition of technogenic raw materials. In the future, physical, mechanical, and technological studies are carried out, according to which a full assessment of the quality of the waste is given in accordance with the requirements of regulatory documents for materials for road construction.

The prospects for the use of specific technogenic raw materials largely depend on the similarity of their properties with similar properties of conventional raw materials. According to the sources of formation, coal mining wastes are divided into overburden mine rocks, current enrichment waste, and rocks stored in dumps. To develop recommendations on the use of mine rocks, it is enough for each species to determine the following main parameters: source of formation; lithological and mineral composition; degree of coalification of organic matter; content of carbon, sulfur, iron, silicon, aluminum, calcium, and magnesium compounds; and also its plasticity. The burnt rocks of mine dumps are specific, insufficiently studied raw materials. Therefore, the recommended parameters can be used only for an approximate assessment of burnt rocks.

The presence of clastic material in burnt rocks, bonded with perlitic and siliceous natural cement to a state of conglomerate, ensures their high water resistance and mechanical strength, reinforced by rock pressure and heat treatment. Under natural conditions, compaction and formation of rocklike (lithified) rock occurred, the physical and mechanical properties of which depend on the degree of metamorphism and other characteristics. Along with dense, well-burnt material, unburned weak rocks, as well as coarse-grained slag, sintered with pieces of rock, can also lie in the dump. Along with large inclusions, dumps contain dusty ash fines resulting from the burning of shales. The quality of burnt rock in the dump can be determined by external inspection by the density of the structure, color, and nature of the surface of the fracture of its pieces. A nodule-like and roughish or greasy to the touch surface in a fracture indicates a poor quality of material with a low bulk density and high water absorption. Well-fired burnt rocks of homogeneous composition, with a dense structure and a smooth stonelike surface in a fracture from dumps of anthracite coals, are distinguished by increased mechanical strength, resistance to abrasion, and temperature and atmospheric influences.

The characteristic of the natural granulometric composition of the burnt rocks is very relative. The grain composition largely depends on the characteristics of the original rocks. The rocks discharged from the mines have a grain size of up to 250 mm, and individual pieces reach 350–450 mm. The stronger the rock, the larger and more massive the calcined material obtained, which is slightly subjected to grinding as a result of firing, sliding off the slopes, and pressure of the overlying layers. Self-firing of fine-grained sandstones and sandstone shales gives the most durable burnt rocks, consisting of large pieces. The fine fraction is obtained from finely and coarse-grained clay and carbonaceous shales, carbonate and glandular minerals having lower mechanical strength.

The degree of burning of burnt rocks can be judged by color, strength, thermal analysis data, as well as water absorption, similar to assessing the degree of sintering of a ceramic crock. Burnt rocks with water absorption in the range of 5–10% are moderately burnt, while water absorption of less than 5% is well burnt.

The physical and mechanical characteristics of rocks of the same type are sometimes different even within the same coal basin and fluctuate over a wide range. The average values of the characteristics of dump burnt rocks of the East Donbass (Russia) are as follows:

(i) True density, 2.2–2.8 g/cm^3
(ii) Bulk density, in pieces 1400–2500 kg/m^3 and crushed 800–1400 kg/m^3
(iii) Porosity, 3.6–32.5%
(iv) Voidness, 22–45%
(v) Water absorption, 0.5–17.5%
(vi) Ultimate compressive strength, 10–100 MPa and higher
(vii) Frost resistance, 10–100 number of cycles
(viii) Abrasion (mass loss), 4.6–19.7%

The most homogeneous in composition and physico-mechanical properties are burnt rocks formed from mine rocks after anthracite mining. Their true density is 2.1–2.7 g/cm^3, and water absorption is from 0.5 to 4%. Burnt rocks formed during firing of coal mining are characterized by a large range of changes in density, strength, and water absorption.

4.2.3 *Chemical Composition of the Burnt Rocks of Mine Dumps*

The chemical composition of burnt mine rocks to a significant degree determines the direction of their possible use. Table 4.1 shows data on the chemical composition of the rock masses of averaged samples taken from some dumps located in the Rostov region (Russia). The data on the chemical composition of the dump of various lithological types are presented: clayey, sandy-clayey, sandy, and calcareous-sandy-clayey. The chemical composition of clay dumps, the most common in the

Table 4.1 Chemical composition of the burnt rocks of mine dumps (mass % per dry matter)

Mine	SiO_2	Al_2O_3	Fe_2O_3	CaO	MgO	TiO_2	P_2O_5	$\Sigma(K_2O + Na_2O)$	SO_3 total	Loss on ignition
Anthracite	58.18	21.80	6.30	1.06	1.15	1.02	0.15	3.07	2.35	4.85
Burgustinskaya	56.65	20.53	6.98	0.92	1.24	1.17	0.21	4.30	3.00	4.93
Vorovskiy	55.58	22.30	6.56	1.02	1.47	1.08	0.14	4.11	2.37	5.32
Komsomolskaya Pravda newspaper	57.49	20.95	6.02	1.61	1.82	1.12	0.17	2.90	1.96	5.90
Mayskay	58.98	19.20	6.85	1.28	2.15	1.01	0.27	3.63	2.65	3.90
Nesvetaevskaya	56.58	19.92	7.04	1.98	1.22	1.07	0.33	3.21	3.15	5.47
Petrovsky	59.03	22.31	6.40	1.12	1.46	1.03	0.35	3.07	1.87	3.27
Rostovskaya	56.83	20.05	6.05	2.13	1.15	1.22	0.17	3.44	3.83	5.08

Rostov region, is quite similar in the content of the main rock-forming oxides. No significant fluctuations in the chemical composition of the rocks were found. The differences are reduced mainly to fluctuations in the content of iron oxides and alkali metals. The chemical composition of the rocks is represented mainly by oxides of silicon, aluminum, and iron. The content of acid oxides is more than 70%, and the rocks can be characterized as siliceous-alumina. The reduction of iron oxide in burnt-out rocks (dump of the Vorovskiy mine) and the high content of sulfur sulfate compounds in the rock mass of the dumps of the Nesvetaevskaya and Rostovskaya mines are characteristics. An important characteristic of rocks that affects the direction of their use is the presence of sulfur. According to this indicator, the rocks are classified as medium sulfur (up to 3%) and sulfur (up to 4%). By the content of organic carbon (0.5–3.3%), the rocks are low carbon. The rocks have high chemical resistance in relation to mineral acids, 95–99%, and to alkalis, 65.5–77.8%.

The chemical and mineral compositions of the rocks are interconnected. The high content of silica is explained by its presence in free form in the form of quartz, as well as bound in aluminosilicates of various compositions (kaolinite, mica, feldspars, etc.). Iron in the rocks is represented by disulfide (pyrite, marcasite), siderite, or a mixture of these minerals. The presence of alkalis can be explained by the decomposition of feldspar, mica, and sulfur trioxide due to the decomposition of sulfates and the oxidation of pyrite. Aluminum, like silicon, is concentrated in the form of a variety of inorganic oxygen compounds.

According to the data of chemical analysis, it is possible to calculate a number of modules reflecting the petrological characteristics and activity of the rocks, silicate, alumina, siliceous, and clay-ferruginous (activity modulus), and evaluate the degree of refinement, weathering, clayness, iron content, and other rock features. The studied rocks have a high degree of variability, the degree of iron content is negligible, and the degree of weathering is increased. Clay constituents prevail in the mineral composition. Rocks of dumps of mines "Burgustinskaya," "Mayskaya," and "Komsomolskaya Pravda newspaper" are sandy-clay (the content of sand in the dumps is over 10%), and the dump rocks of the Almaznaya mine are calcareous-clay. The clay part is represented mainly by hydromica mixed with kaolinite. According to the basicity modulus, which takes into account the ratio of acidic and basic oxides, the rocks are acidic, and the basicity modulus is 0.03–0.069. According to the size of the clay-iron module (0.4–0.6), the studied rocks belong to group I (mudstones are active and highly active) [77, 78].

A feature of the chemical composition of the burnt rocks of the dumps of calcareous-sandy-clay composition is the high content of calcium and magnesium oxides. This is characteristic of mine rocks of dumps of Gukovo-Zverevsky and Kamensko-Gundorovsky districts of Rostov region (Russia).

The remains of coal and carbonaceous mudstones are present in the burnt rocks. The carbon content is estimated by the concentration of organic carbon. The amount of organic carbon can range from 1.2% to 7.5%. Significant variation is noted in the magnitude of the loss on ignition. High losses on ignition (more than 3%) of burnt rocks indicate incomplete burnout of organic substances.

The rocks of the studied dumps contain readily soluble salts, amorphous silica, and free calcium oxide. Their content is small. The mineral part of water extracts consists of various water-soluble salts, among which bicarbonates, calcium, and magnesium sulfates predominate. Calcium bicarbonates prevail in the water extracts of the rock mass of calcareous dumps.

In burnt rocks, in addition to macronutrients, trace elements are also present. The microelement composition of burnt rocks is less studied than the composition of macroelements. This is mainly due to the difficulties in analyzing the content of trace elements. At the same time, in assessing the possible areas of use of burnt rocks, one should take into account these data from the point of view of the prospects of extracting valuable trace elements. So, to determine the agrochemical properties of the rock mass, it is necessary to determine the content of manganese, copper, zinc, molybdenum, cobalt, boron, and iodine. In addition to identifying these elements, a complete semiquantitative spectral analysis of the rock mass is performed. It should be borne in mind that the rare elements (germanium, uranium, gallium, vanadium, and rhenium) may be of industrial interest.

In published works [124, 125], the content, distribution of trace elements, identification of general patterns, their accumulation in solid fossil fuels, and dump or by-products are considered in sufficient detail.

4.2.4 Mineralogical Composition of the Burnt Rocks of Mine Dumps

The mineralogical composition of the rocks of mine dumps is determined by the mineral composition of the carbon-bearing rocks and is closely related to the subsequent impact of various processes leading to its change. The rocks of coal deposits are composed mainly of sedimentary rocks: sandstones, mudstones, and siltstones, and less often limestones. Sandstone is sand cemented by silicic acid, limestone, or clay. Mudstone is compacted, dehydrated, and cemented clay with particle sizes less than 0.01 mm. Siltstone is a cemented rock intermediate between sandstone and mudstone with a particle size of 0.01–0.1 mm. The mineral composition of sandstone, mudstone, and siltstone is represented mainly by quartz, fragments of feldspar, kaolinite, hydromica and montmorillonite mixed with chlorite, and hydroxides of iron, aluminum, and silicon. Limestone consists mainly of calcite mixed with dolomite, clay, and sand particles. During prolonged firing, the initial carbohydrate rocks undergo various changes to one degree or another and acquire properties that are different from the properties of the initial rocks.

Thus, the high-ash rock mass stored in dumps is represented by a mixture of carbonaceous rocks, siltstones, mudstones, sandstones, and limestones. The ratios between minerals vary over a fairly wide range. According to the lithological composition, depending on the prevailing type of rocks, the following types of dumps are distinguished [130, 147]:

(i) Clay, consisting of mudstones and siltstones with a sandstone content of up to 10% of the total volume of the rock
(ii) Sandy-clayey (sandstones 10–30%)
(iii) Sandy (sandstones more than 30%)
(iv) Calcareous-clayey (sandstones up to 10%)
(v) Calcareous-sandy-clayey (sandstones 10–30%)

On the whole territory of the East Donbas (Russia), the clay type of dumps prevails. The composition of the waste rocks is unstable. The most homogeneous in lithological composition are dumps working out one coal layer. When mining several coal layers, mine dumps are characterized by a polymineral lithological composition.

By the lithological composition of the clay part of the rocks, argillite, carbonaceous-argillite, mudstone-siltstone, and siltstone types of dumps are distinguished. According to the lithological composition of the rock mass, mine dumps of the Rostov region (Russia) are divided into five types.

Among burning and burnt dumps with a large stock of rocks, we note the following:

(i) *Clay*: dumps of the mines “South” and “Nesvetaevskaya”; mines “Petrovsky,” No. 26, “Anthracite,” and “Central” (Gukovo-Zverevsky district)
(ii) *Sandy-clayey*: dumps of the mines “Mayskay,” “October Revolution,” “Komsomolskaya Pravda newspaper,” “Burgustinskaya,” etc. (Gukovo-Zverevsky district)
(iii) *Sandy*: dumps of the mines “Gornyatskaya-Yuzhnaya,” “Belokalitvenskaya,” mines No. 4 and No. 11 (Belokalitvensky district)
(iv) *Calcareous-clay*: dumps of the mines “Oktyabrskaya,” “Almaznaya,” “Izvarinskaya,” etc. (Gukovo-Zverevsky district)
(v) *Calcareous-sandy-clayey*: dumps of the mines “Gundorovskaya,” “Donetskaya” (Kamensko-Gundorovsky district)

Limestones are found in the dumps of mines of the Kamensko-Gundorovsky and Gukovo-Zverevsky geological and industrial regions. Limestones do not occur in sandy dumps.

In the rocks, four main types of textures are distinguished:

(i) *DENSE*, characteristic of slightly modified siltstone and sand varieties
(ii) *Porous* (spongy) formed during the burning of carbonaceous particles in various lithological differences
(iii) *Slag-like*, resulting from the fusion of clay with pyrite
(iv) *Heterogeneous*, which are conglomerates of separate fragments with different textures

In the dump, the rock mass is a multicomponent system, the burning of which is accompanied by a parallel-sequential flow of many reactions leading to the formation of many end products depending on the temperature and ratio of various components of the system.

During self-firing of mine rocks, complex phase transformations occur [77]:

(i) Removal of adsorbed moisture (120–160 °C)
(ii) Destruction (250–500 °C)
(iii) Oxidation and burnout of organic matter (300–1000 °C)
(iv) Dissociation and oxidation of iron sulfides (300–500 °C)
(v) Dissociation of iron carbonates (500–700 °C), calcium and magnesium (700–850 °C)
(vi) Stepwise dehydration and amorphization of clay minerals (550–900 °C)
(vii) Crystallization of newly formed compounds (mullite, sillimanite, hematite, spinel), as well as their interaction with each other or with oxides to form silicates and aluminosilicates (980–1200 °C)
(viii) Melt formation (glassy phase, mainly melinite composition)
(ix) Crystallization of new phases: magnetite, cristobalite, corundum (950–1200 °C)

During physical and chemical weathering of rocks, a change in their size occurs due to the dispersion and destruction of chemically unstable minerals. As a result of thermal action of varying intensity and uneven distribution of the focus of combustion, rocks of various degrees of variability are formed: calcined, burnt, remelted, and rocks calcined in a medium close to recovery one.

The color of the calcined rocks differs from the original ones by the prevalence of reddish-brown tones. In calcined mudstones and siltstones, mainly clay substance is subjected to thermal effects: it is partially amorphized and covered with films of iron hydroxides. In the sandstones, mostly debris is affected. The texture and structural features of sandstones and siltstones remain unchanged.

The greatest changes in the mineral composition of the rock mass during self-firing occur during the rocks remelting up to the formation of new types of rocks. Macroscopically they are dense masses from reddish-brown to brick-red in color with dark brown and purple hues. In burnt rocks, two groups of minerals can be distinguished: the minerals of the initial rock and newly formed [71, 72]. The first group includes quartz, feldspars, mica, partially hematite, rock fragments, hydromica, and kaolinite. The second group includes mullite, forsterite, cordierite, spinel, tridymite, sillimanite, andalusite, glass, hematite, etc. The most profound changes occur in the mineral composition of mudstones. Their composition is dominated by clay particles amorphized by firing, coated with iron hydroxides and silty particles of silica, almost all changed: cloudy and fissured. The texture can be defined as bubble-layered spotted.

Remelted rocks have a dark gray color with areas of a brownish tint and cavernous and porous textures. They are composed of aggregates of solid anhydrous silicates and oxides, and fragments of melted rocks form breccia-like lumps resembling sintered and welded agglomerate tuffs with a slag texture.

Dense rocks of black color with a characteristic metallic luster were also found in dumps: “black blocks” [43]. They formed in the inner parts of dumps when the rocks were calcined in a medium close to recovery one. These rocks contain a significant amount of carbon (15–20%) in the form of finely dispersed graphite and fragments of former coal.

Changes that occur with rocks during thermal treatment were recorded by X-ray and differential thermal analysis of siltstones and mudstones [115]. In the initial sample of siltstone (Fig. 4.1a) the lines of diffraction reflection $d/n = 0.238$, 0.357, and 0.714 nm are characteristic of kaolinite, and the peak intensities are $d/n = 0.331$ and 0.498 nm for hydromica. Strong lines of montmorillonite 1.105 nm and quartz ($d/n = 0.333$; 0.425 nm) are noted.

The diffraction pattern of a sample of aleurolite calcined at 550 °C (Fig. 4.2b) is characterized by hydromica ($d/n = 0.329$ and 0.495 nm) and quartz with a reflection of 0.334 nm. The lines of kaolinite disappear, which indicates its dehydration.

For mudstones (Fig. 4.2), the presence of kaolinite can be judged along the lines of 0.238, 0.357, and 0.714 nm, and hydromica is defined by peak intensity of 0.331, 0.495, and 1.078 nm. The presence of quartz is confirmed by reflexes of 0.425 and 0.333 nm, and siderite corresponds to 0.281 nm. On the diffraction pattern, taken from a sample of calcined quartz, mica and chlorite are observed.

Siltstones and mudstones are characterized by thermal effects associated with the release of hygroscopic and constitutional moisture at 145–180 °C and exothermic effect at 400–500 °C, indicating the onset of burning of organic matter. The dehydration of kaolinite and the formation of metakaolinite correspond to the endothermic effect at 580–600 °C; decomposition of organic matter occurs at 650–680 °C.

The endothermic effect at 740–780 °C is characteristic of hydromica and mixed-layer minerals; at 860–980 °C, their structure is transformed. The exothermic effect at 980–1100 °C indicates the formation of new crystalline phases: mullite and spinel.

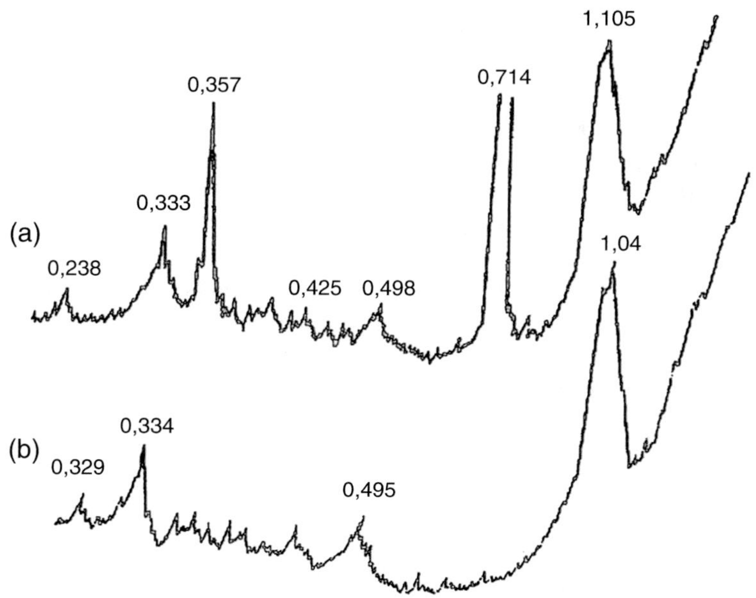

Fig. 4.1 Diffraction pattern of siltstone: (**a**) original siltstone; (**b**) siltstone calcined at 550 °C

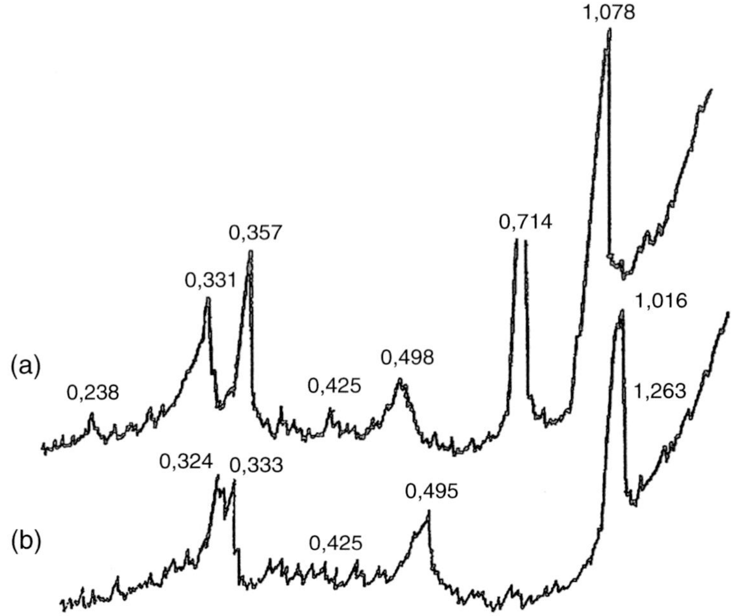

Fig. 4.2 Diffraction pattern of mudstone: (**a**) original mudstone; (**b**) mudstone calcined at 550 °C

The multimineral composition of burnt rocks requires the use of comprehensive studies to reliably assess their material composition, quality indicators, and technological properties. These studies include lithological and mineralogical methods, X-ray diffraction and thermal analysis, and transmission and scanning electron microscopy, with simultaneous characterization of the chemical composition and physico-mechanical properties. Their use provides almost complete information about burnt rocks and allows one to choose the optimal direction of their utilization.

The cross section, shown in Fig. 4.3a, is made of burnt ocher yellow with spots of reddish-brown color. The rock is dense, strong, and heavy, and is divided into tiles, on the separate planes of ocher grease. The structure is sandstone, medium grained. The composition of the sandstone is polymineral. Among the detrital part of the rock, quartz predominates; there are also fragments of siliceous rocks, feldspars, mica, chlorite, carbonates, accessors (zircon, etc.), and ore minerals. Grains of 0.25–0.50 mm in size prevail. The largest grains reach 0.60 mm. The nature of cementation is contact-pore. The composition of the cement is hydromica and clay, and sometimes siliceous. The clay component of the cement is represented by kaolinite with relics of the "collar structure." Debris and cement are largely altered by self-burning processes. The grains of quartz are opaque, fissured, and contain numerous inclusions, mainly ore. At the edges of the grains melted, an anomaly of optical properties is observed. The consequence of firing is also the presence of a large number of altered ore minerals present in the form of individual grains, films, pigment spots, etc.

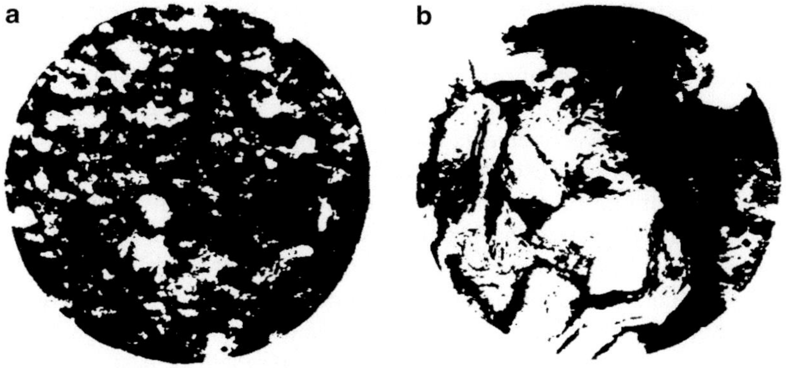

Fig. 4.3 Microphotographs of burnt rock (×140, Nicoli N): (**a**) sandstone fine-grained structure; (**b**) silt-pelitic structure

The cross section shown in Fig. 4.3b consists of brick-red, porous, light rock with a thin scaly, sometimes leafy, characteristic of mudstones. The microstructure of the rock is porous. Pores occupy about 25–30% of the area. The pore shape is irregular, the edges are uneven and torn, and the average pore size is 0.1–0.2 mm. With one Nicole, it can be seen that the rock consists of brownish lumps of altered pelite; in reflected light they have a yellow color with a reddish tint; and in some places point inclusions of red hematite are observed. Quartz is present in a significant amount in the form of small angular grains with a size of 0.003–0.007 mm. In mudstone, oxidized carbonate is unevenly distributed, corresponding in composition to siderite or ankerite.

The study of the material composition of burnt rocks in immersion makes it possible to determine quartz, feldspar, and clay pelitic products altered during self-firing in their composition. Clay substance and iron hydroxides in the form of films and shirts cover aggregate grains. Clay is sometimes partially decomposed and replaced by iron hydroxides. It was established that, as a result of thermal effects, quartz grains sometimes undergo modifications, and their contours are blurred.

Microscopic studies of the rock make it possible to determine argillite and siltstone varieties of heterogeneous color, dense, thinly shale structure. They often alternate with interlayers enriched with carbonaceous clay. Shattered glass fragments, colorless and colored, have a refractive index of 1.48, 1.55, and 1.67, characteristic of feldspar and clay glasses.

The presence of large amounts of iron in crystalline and colloidal form, as well as the complex polymineral composition of burnt rocks, makes it difficult to study their material composition. In this regard, the burnt rock was subjected to various types of chemical treatment, for example, heated aqueous solutions of sulfuric, hydrochloric, and phosphoric acids. The most effective was the treatment with a 5% aqueous solution of oxalic acid in the presence of aluminum. When boiling burnt rocks in this solution, the glandular films dissolve and oxalic iron is deposited on the aluminum surface. With this treatment, partial dispersion of the burnt rocks occurs and removal of a significant portion of the iron. The above image (see Fig. 4.4) of the rock after processing confirms the expected result. Films of iron hydroxides are mostly removed from the surface of the rock particles. The mineral composition of the pelitic and siltstone constituents was studied on rock mass samples of the mine dump No. 26. It was established that the secondary minerals formed in the heavy fraction of the silt constituent formed during the burning of the dump and subsequent weathering of the rocks: hematite, magnetite, limonite, glass, barite, etc. These minerals are also contained in the pelitic component (up to 16%). The bulk of the pelitic component is represented by hydromica (up to 60%) with an admixture of kaolinite and finely divided rock fragments (up to 20%).

The light fraction of the silt constituent is represented in basic quartz (up to 50%). The main rock-forming minerals of the rock mass are quartz, hydromica, and kaolin. A distinctive feature of the burnt rocks of the mine dump No. 26 is the high content of iron hydroxides (hematite, limonite).

X-ray analysis of burnt rocks is very complicated, firstly, by the variety of mineral composition, and secondly, by the presence of some minerals in a finely dispersed state. As a result, the lines in the X-ray diffraction patterns are diffuse and weak in intensity. Figure 4.5 shows X-ray diffraction patterns of burnt rocks of some dumps. Judging by the values of interplanar distances, the rock contains the following basic crystalline minerals:

(i) Quartz and its modifications: $d = 3.35, 2.29, 2.12, 1.82, 1.67$
(ii) Cristobalite with $d = 4.05, 2.52, 2.21$
(iii) Iron in the form of hematite with $d = 2.69$ and 1.69 and magnetite with $d = 2.53$ and 2.97
(iv) Clay substance with $d = 2.70, 2.46, 1.97$; mullite with $d = 2.20, 1.69, 1.59$

As minor minerals, kaolinite (d = 7.14; 4.50; 4.30; 1.54), feldspars (d = 3.66; 3.18; 2.88), sulfur (d = 2, 85; 2.12), and others are present. The presence of a notice-

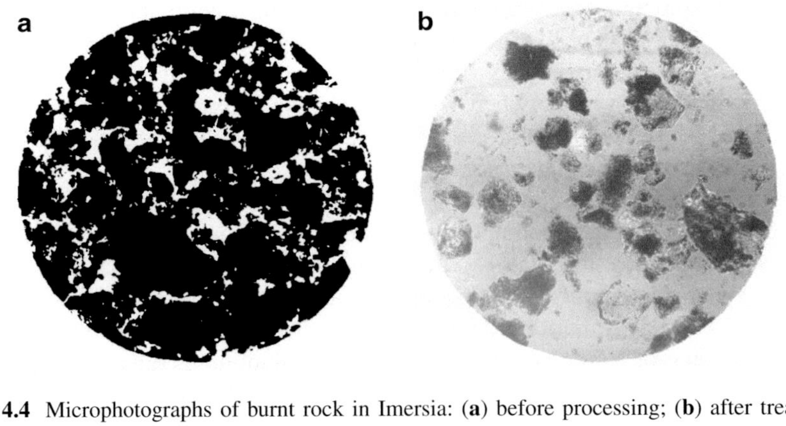

Fig. 4.4 Microphotographs of burnt rock in Imersia: (**a**) before processing; (**b**) after treatment with oxalic acid

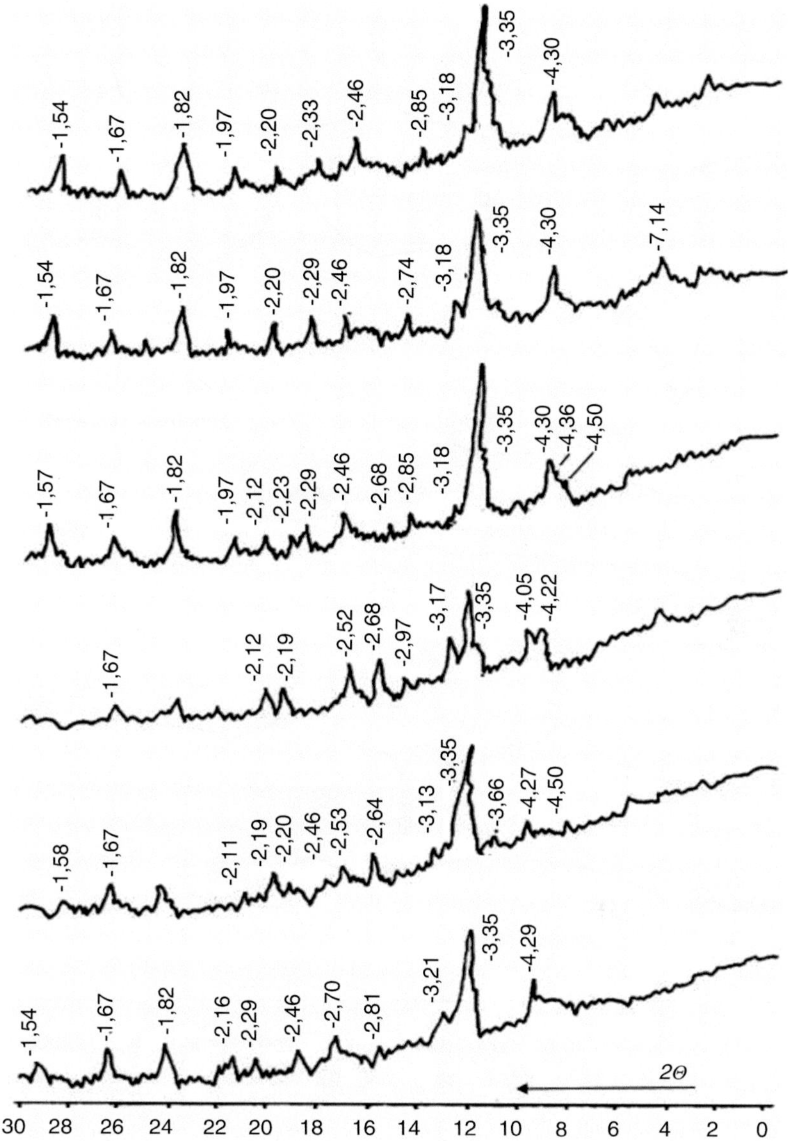

Fig. 4.5 Radiographs of the burnt rocks of mine dumps: 1—"Petrovsky"; 2—No. 26; 3—"Burgustinskaya" No. 3; 4—"Maiskaya"; 5—"Central"; 6—"Anthracite"

able background indicates existence of amorphous and finely dispersed colloids in the burnt rock.

On thermograms (see Fig. 4.6) of the studied burnt rocks, endothermic effects are observed at a temperature as follows:

(i) 170–200 °C—removal of hygroscopic moisture
(ii) 460–550 °C—dehydration of clay minerals and removal of bound water

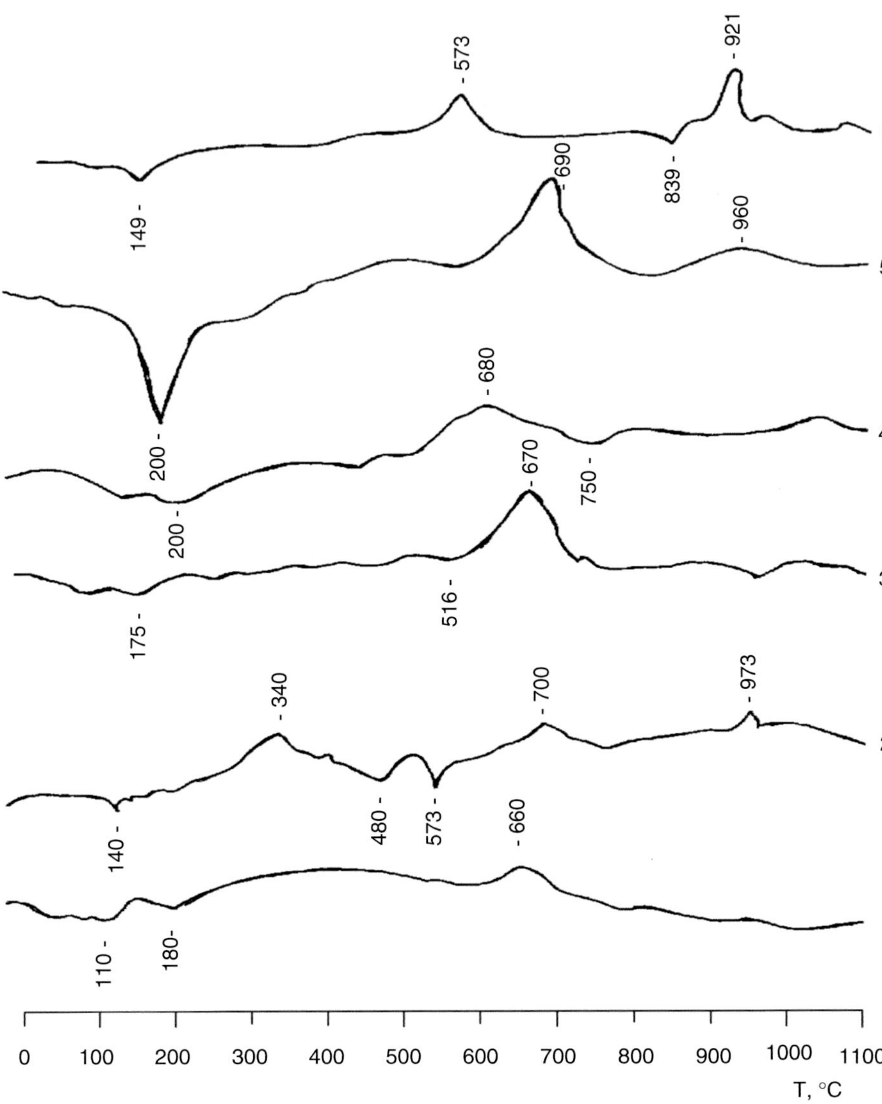

Fig. 4.6 Thermograms of the burnt rocks of mine dumps (designations are the same as in Fig. 4.5)

Exothermic effects at a temperature as follows:

(i) 130–280 °C—oxidation of glandular minerals
(ii) 573–1000 °C—polymorphic transformations of quartz, burning out of carbonaceous substances, and decomposition of calcium and magnesium carbonates

According to thermal analysis, we can conclude that the content of carbonaceous substances in the rocks of the Burgustinskaya, Tsentralnaya, and Nesvetayskaya mines is slightly increased, as well as the absence of effects characteristic of clay

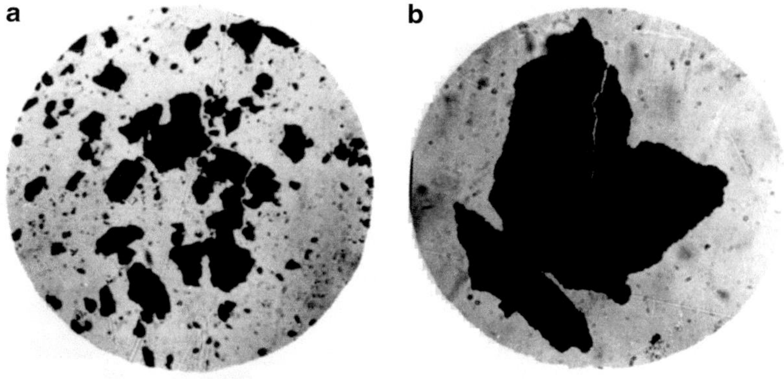

Fig. 4.7 Microphotographs of carbonaceous impurities: (**a**) ×63, Nicoli H; (**b**) ×140 Nicoli H

minerals in burnt rocks. Peaks in the thermograms appear as a result of processes occurring in the part of the rock that underwent insufficient firing in the dump or was burnt at a lower temperature. The insignificant peaks show that there are few such rocks. Well-burnt rocks are thermally inert, and on the differential thermal analysis curves they almost do not show thermal effects. The amount of carbonaceous matter and its modification species has a significant effect on the activity of burnt rocks and the physical and mechanical properties (especially durability) of products based on them. In this regard, a complex investigation was carried out on a carbonaceous substance isolated separately by boiling.

Figure 4.7 shows micrographs of a carbonaceous substance, which is represented by particles of various sizes and shapes. Part of the grains is in a finely dispersed state; there are spherical particles with a diameter of 0.02–0.19 mm. All particles are opaque, painted in black; in reflected light they have a gray color and a metallic luster. The bulk of the irregularly shaped particles are semicoke and coke residues. Translucent at the edges or completely transparent particles, apparently, represent aggregated clay minerals with sprayed particles of dispersed carbon. The presence of carbon is explained by the decomposition during thermal processes of coal and carbon monoxide present in the heaps. Semicoke varieties of carbonaceous impurities: the product of the intermediate coking process (mesophase) is represented by spherical particles in the image.

Carbon impurities in the form of graphitized opaque constituents, close to graphite, partially transforming into semicoke varieties, were detected by X-ray diffraction (see Fig. 4.8). The presence of graphitized carbonaceous matter can be judged by the interplanar spacings: d = 2.11, 2.28, 3033, and 3.71. On the carbonogram, thermogram presented in Fig. 4.9, peak characteristics of the processes of graphitization and coking of coal are recorded.

Thus, the carbonaceous impurities present in the burnt rock consist of coking products (semicoke and coke residues), colloidal carbon sprayed onto the surface of the mineral particles, and graphitized carbonaceous matter. Moreover, sulfide formations, namely pyrites (pyrite, marcasite, etc.), are present in the rock. During the oxidation of the latter, the rock mass is enriched with sulfur.

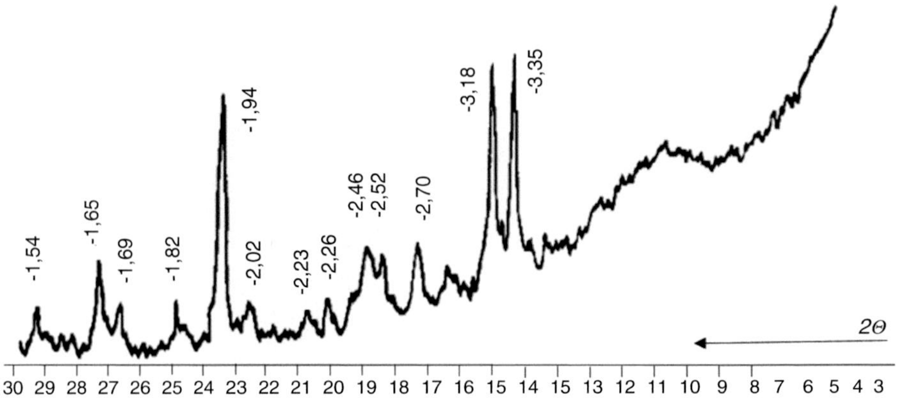

Fig. 4.8 Radiograph of the carbonaceous substance of burnt mine rock

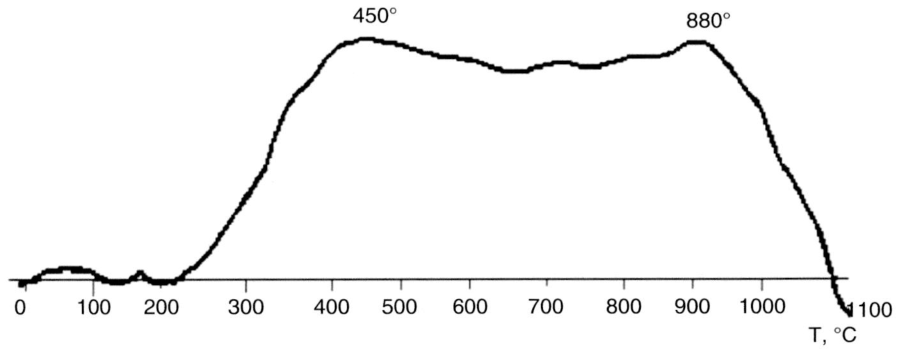

Fig. 4.9 Thermogram of carbonaceous matter by burnt mine rock

A comprehensive study of the burnt rocks of dumps shows that their mineral composition is almost identical. The bulk of the rocks is composed of opaque particles of an amorphized clay-fired substance coated with films of iron hydroxides such as hydrohematite. Quartz is present in significant amounts in the form of pulverized, siltstone, and sand grains. There are feldspar grains, and individual grains of ore minerals. Sulfur and carbonaceous inclusions are present. In the description of thin sections, it was found that all minerals were significantly altered during the firing process: quartz grains were cloudy, cracked, and fused along the edges, and the clay substance was amorphized. When considering rocks in immersion, quartz, feldspars, mica, carbonates, ore minerals, and glass fragments, transparent and painted in yellow and occasionally reddish-brown, are identified. The data of X-ray and thermal analyzers confirm the diversity of the mineralogical composition of the studied burnt rocks of mine dumps.

4.3 Ash-slag Waste

4.3.1 Methods of Ash-slag Waste Selection

Solid fuel is burnt both in the form of pieces and in the form of coal dust. The ash-slag of TPPs is the remainder of solid fuel combustion. They are the products of high-temperature (up to 1200–1700 °C) processing of the mineral, unburnt part of the coals. Lump fuel is used only in small enterprises. The bulk of the ash-slag is formed at thermal power plants from pulverized coal. At the same time, two types of waste are obtained in chamber furnaces: fly ash and slag. By the type of fuel burnt, anthracite, coal, and brown ashes are distinguished; the chemical composition of the ash is divided into acidic and basic. It is advisable to use ash-slag with a high content of calcium and magnesium oxide as a raw material for the production of binders, and low calcium as active fillers and mineral additives in the production of building materials and in road construction.

Various ash collectors are used to extract ash from power plants: cyclones and battery cyclones, wet ash collectors (scrubbers), and electrostatic precipitators. Cyclones well retain large fractions of ash, and badly the smallest ones. They capture 50–80% of the ash. To increase the efficiency of gas purification, cyclone elements are reduced in diameter and combined into battery cyclones. The degree of ash collection in scrubbers is 80–95%. Flue gases are supplied to them tangentially, as well as to cyclones. The inner walls of the scrubber are washed with water. Particles of ash, thrown to the walls by centrifugal force, flow down with water down into the hydraulic removal system.

A very high degree of gas purification from ash (99.0–99.5%) is provided by electrostatic precipitators. In the chamber of the electrostatic precipitator, precipitation and corona electrodes are installed at a strictly defined distance from each other; when a high voltage is applied, a corona discharge arises. Particles of ash are charged and deposited on the electrodes, and the purified gas is discharged to the outside. The ash settled on the electrodes is continuously or periodically shaken off and enters the hopper of the electrostatic precipitator.

Fly ash trapped in cyclones and electrostatic precipitators consists of particles that differ in size and density, due to the difference in their phase and material composition. Thus, an increase in density is usually caused by the content of calcium and iron oxides FeO + Fe_2O_3 in the particles. The latter form the magnetic component of the ash. Iron oxides are concentrated mainly in large fractions of ash. The finest grained ash with a maximum glass-phase content can be obtained in multicyclones and electrostatic precipitators.

From the bunkers of the ash collectors, ash is removed in two ways: hydraulic and pneumatic. Most power plants use the hydraulic method, in which the ash-slag mixture removed from the boiler houses is hydraulically directed to specially equipped areas of the terrain, namely ash dumps. When the pulp leaves the ash duct, particles of different masses under the influence of gravitational forces are deposited at different distances from the inlet of the ash duct. As a result, a large heterogeneity is observed in the area of the dump according to the granulometric composition and content of coked particles. Larger particles (>0.25 mm) and heavier particles are concentrated near the place of pulp discharge, and small particles (<0.25 mm) settle on the periphery. As a result of the displacement of the pulp discharge site, these zones are randomly distributed over the territory of the ash dump, which, with the subsequent development of the dump, enhances the heterogeneity of the consumed raw materials. The heterogeneity in grain composition also causes some heterogeneity in the chemical composition, which degrades the properties of ash-slag waste from hydraulic dumps as raw materials for the production of building materials and complicates their use.

The simplest and most effective way of averaging hydraulic dumping ashes is their multiple transshipment. Averaging occurs during the preparation of slag mixture at hydraulic dumps into cones, where it dries to a moisture content of 20–25%, and then it is loaded into hopper cars and unloaded at an open warehouse after delivery to the place of consumption. As a result of the operations performed, the homogeneity of the grain composition of ash-slag waste increases, and dusting during transportation and freezing in winter are reduced.

A more progressive method for removing ash is pneumatic. According to the standard scheme of dry ash selection from ash bins, under an electrostatic precipitator, it is pneumatically fed into a vacuum ash pipe, a precipitation chamber with a filter, and a storage hopper. From the storage bunker, ash is pumped into a silo. With the help of compressed air supplied through the pipeline, the ash is loaded into railway hopper cars or auto trucks. With the dry method of collecting ash, there are possibilities to increase its homogeneity.

4.3.2 Physical Properties of Ash-slag Waste

When assessing the physico-mechanical properties of ash, slag, and their mixture, it is necessary to take into account particle size distribution, composition uniformity, particle shape and structure, specific surface area, and true and bulk density. Ash and slag are composed of particles differing in both true and bulk density, characterizing the porosity and particle size distribution of the waste. These indicators are related to the chemical and mineralogical composition. So, in larger fractions, an increased content of iron or dense minerals is noted. The properties of ash are influenced by the shape and surface structure of its particles. Dusty fuel in boiler furnaces burns out for a short time at a temperature of 1000–1700 °C. Due to the heterogeneity of the composition of individual particles of the fuel mixture and the inability to contact dust particles that are burnt in suspension in a gas stream, mineral formation reactions occur in volumes of individual particles [62]. This leads to the ash of solid fuel being represented by particles of various shapes and sizes. Numerous studies of ash have established a variety of particle shapes (Fig. 4.10). Particles of two types should be distinguished in ashes: homogeneous, which include spherical, acute-angled, filiform, and aggregated, consisting of several grains. Spherical particles are found in the ashes of all power plants and have a color

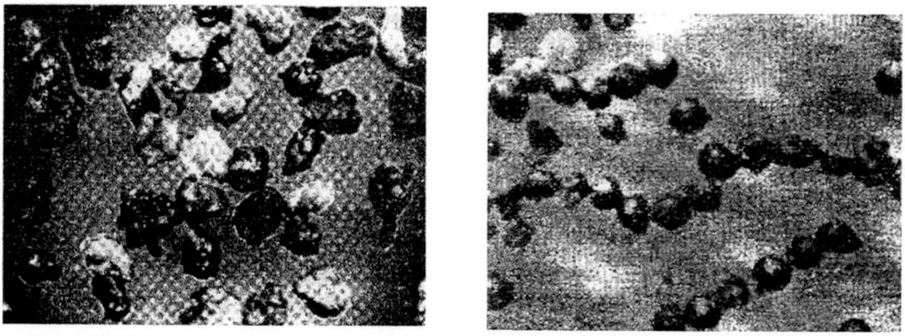

(a) (b)

Fig. 4.10 Micrographs of ash particles: (**a**) particle size of 0.08 mm; (**b**) particle size of 0.063 mm

from light to black with a metallic sheen. Glassy particles of acute-angled shape are much less common. With a decrease in particle size, their shape changes. The smaller the ash fraction, the more it contains melted particles. Most of the completely melted particles upon cooling form glass droplets, which often contain bubbles that are closed or less often opened at the surface of the droplet. Large fractions contain more particles of irregular shape, angular, sometimes having a porous structure.

Glassy particles of acute-angled shape are much less common. With a decrease in particle size, the shape changes. The smaller the ash fraction, the more it contains melted particles. Most of the completely melted particles upon cooling form glass droplets, which often contain bubbles that are closed or less often opened at the surface of the droplet. Large fractions contain more particles of irregular shape, angular, sometimes having a porous structure. Ashes generally have a spherical particle shape and a smooth, vitrified surface shape. The most homogeneous particles are all composed of glass. There are also particles, the inner part of which consists of minute mineral and coke grains. There are also hollow balls formed as a result of the expansion of glass at the time of formation of particles. Large fractions of ash contain aggregate particles resulting from the sintering of many small grains. They are heterogeneous and brittle, and easily crushed.

Fly ash is a finely dispersed material; the size of its part varies from 3–5 to 100–150 microns. The quantity of larger particles usually does not exceed 10–15%. An important characteristic that determines the properties of ash and the possible directions of its use is the specific surface. It can range from 200 to 700 m^2/kg. The specific surface of the ash is determined by the shape of the particles: the more rounded particles, the smaller the specific surface, ceteris paribus. The true density of individual ash fractions can vary from 2 to 2.5 g/cm^3. Despite the large difference in the true density of individual fractions, usually the average true density of the entire ash sample (including all fractions) for various coals varies within narrow limits: from 1.75 to 2.5 g/cm^3. These fluctuations are often determined by the content of unburnt particles, the type of fuel, and the grain composition of the ashes.

The true density of coke and semicoke particles does not exceed 1.3 g/cm^3. Therefore, aggregated or individual coke particles have a lower bulk density than vitreous. Large fractions of ash usually contain more unburnt fuel than small fractions. This is reflected in their true density. In the event that unburnt fuel is almost completely absent in the ash, the true density of various fractions can vary a little. In the presence of unburnt fuel, large ash fractions have a lower true density.

The bulk density of ash varies between 600 and 1300 kg/m^3. Depending on the volumetric weight in the bulk state, ash can be divided into particularly light (up to 500 kg/m^3), light (500–700 kg/m^3), medium (700–1000 kg/m^3), and heavy (more than 1000 kg/m^3). Bulk density of slag mixture is 1100–1600 kg/m^3.

The slag mixture contains fly ash and slag. The ratio of fly ash and slag is determined by the technology of burning fuel. In furnaces with solid slag removal, 10–20% of ash is transferred to slag, with liquid slag removal up to 20–40%, and in cyclone furnaces up to 85–90% [103]. Ashes captured in cyclones and electrostatic precipitators differ in size and density and have differences in their phase and material composition. According to granulometric composition, ashes are classified into three groups: large, medium, and small, and by homogeneity: monodisperse and polydisperse. The latter prevail. The fine fraction includes particles finer than 20 μm, medium with particles from 20 to 85 μm, and coarse with particles more than 85 μm. In the small fraction, the modulus of fineness is M_f = 2.0, in the middle fraction M_f = 2.0–2.5, and in the large fraction M_f = 2.5–3.0.

The size of the ash cannot fully characterize its dispersion. The reason for this is microporosity of particles. Porosity can be not only closed, but also open, and the total pore volume can reach 60% of the particle volume. The source of pore formation is the processes of dehydration of clay minerals and dissociation of particles of limestone, gypsum, and organic substances. As a result of the occurrence of the gaseous phase, the droplets of the viscous melt swell. The extremely small pore size determines the very large real surface of the ashes. Its specific surface is associated with its properties such as hygroscopicity, water demand, and hydraulic activity.

The dry ash size of the Novocherkasskaya state district power station (Rostov region, Russia) is small, the minimum particle size does not exceed 0.25 mm, and the specific surface area is mainly 300–400 m^2/kg. When ash and slag are removed together, polydisperse ash-slag mixtures enter the dump located near the state district power station. Ash-slag mixtures of various grain composition are accumulated in ash dumps (small, medium, large). The dispersion in the composition is shown in Table 4.2. The modulus of fineness of the slag mixture similar to sand ranges in 1.2–3.7.

True density of ash particles: 2.15–2.47 g/cm^3, slag mixture: 2.3–2.5 g/cm^3, bulk density of ash: 1100–1300 kg/m^3, slag mixture: 1400–1600 kg/m^3; porosity of ash: 20–37%, and slag mixture: 20–50%. The heterogeneity of ash-slag waste in terms

Table 4.2 Granulometric composition of the slag mixture of Novocherkasskaya TPP

Fraction, mm	>10	5–10	2–5	1–2	0.5–1	0.25–0.5	0.125–0.25	0.125–0.4	<0.4
Amount, %	0–35	0–10	15–30	10–15	10–20	0–3.5	4–4.7	20–50	2.5–85

of particle size distribution makes it difficult to use. At the end of the construction, at the ash dump of the slag mixture-processing enterprise, air-conditioned products will be shipped to the consumer.

4.3.3 Chemical Composition of Ash-slag Waste

The chemical composition of ash-slag waste generated during the combustion of ordinary anthracite coal in the Donetsk basin (Ukraine) is mainly represented by silicon and aluminum oxides. Evaluation of the suitability of ash for various purposes is made taking into account its chemical composition. Typically, the chemical composition of the ash is considered as a characteristic of its potential hydration ability. For a more detailed assessment of the quality of ash-slag waste their mineralogical composition and physical and mechanical properties should also be taken into account.

Tables 4.3 and 4.4 show the results of the analysis of samples of ash and ash-slag mixture of Novocherkasskaya TPP, taken and studied at different times. Ash-slag waste consists of free and bound-in chemical compounds, the oxides of silicon, aluminum, iron, calcium, magnesium, potassium, sodium, and sulfur. In the studied waste, calcium oxide contains less than 10%, and acid oxides more than 70%. Free calcium oxide was practically not found in the ash. The content of amorphous silica is no more than 2% and in the form of quartz is no more than 10%. The ash content of organic substances is negligible. The absolute value of losses during calcination is up to 12% for ash, and up to 15% for slag. Carbon inclusions in ash-slag waste are present mainly in the form of coke and semicoke

Table 4.3 Chemical composition of the fly ash of Novocherkasskaya TTP (mass % on dry matter)

Probe	SiO_2	Al_2O_3	Fe_2O_3	CaO	MgO	TiO_2	SO_3 Com.	$\Sigma(K_2O + Na_2O)$	Loss on ignition	Clay-glandular module
1	53.09	20.30	7.32	2.55	2.11	0.93	0.58	3.49	9.18	0.52
2	54.60	21.03	7.54	1.83	2.25	0.87	0.35	3.50	7.37	0.52
3	52.25	21.04	7.91	2.09	2.35	0.93	0.38	3.65	8.81	0.55
4	52.80	23.05	8.00	2.86	1.60	0.97	0.52	2.05	7.66	0.59
5	52.67	22.10	5.44	2.48	1.78	0.91	0.20	2.28	11.4	0.52
6	54.90	21.40	6.13	4.38	1.88	0.93	0.62	2.18	7.68	0.48
7	53.35	20.08	7.51	2.67	2.16	0.93	0.64	1.15	10.2	0.52
8	49.75	23.80	8.18	2.81	2.06	0.96	0.69	2.03	8.61	0.64
9	52.45	21.32	7.45	4.96	2.30	0.99	0.62	2.00	7.35	0.55
10	51.50	21.20	7.37	3.07	1.97	0.84	0.63	2.08	10.4	0.55
Mean value	52.74	21.49	7.28	2.97	2.05	0.93	0.52	2.44	8.87	0.55

Table 4.4 Chemical composition of the ash-slag mixture of Novocherkasskaya TPP (mass% on dry matter)

Probe	SiO_2	Al_2O_3	Fe_2O_3	CaO	MgO	MnO	SO_3 Com.	$\Sigma(K_2O + Na_2O)$	Loss on ignition	Clay-glandular module
1	51.50	26.35	5.35	12.34	0.43	–	0.06	2.77	6.42	0.622
2	48.15	28.25	2.90	5.42	1.56	–	0.70	2.80	10.20	0.647
3	47.94	22.44	9.60	3.55	2.30	0.07	0.79	2.83	9.07	0.689
4	53.50	21.10	9.80	4.34	3.23	0.07	0.98	2.70	4.00	0.577
5	53.00	22.10	9.86	3.60	2.00	0.03	0.70	2.87	3.00	0.634
6	49.80	23.30	5.35	5.85	2.00	0.05	0.25	3.14	9.40	0.575
7	54.48	22.26	4.78	2.81	2.11	0.06	0.44	1.78	11.03	0.496
8	54.30	21.84	5.42	2.93	2.16	0.05	0.44	1.97	10.07	0.502

residues. As a result of metamorphism at high temperatures, incompletely burnt fuel particles turn into coke and semicoke residues. Their resistance to oxidation and durability when exposed to moisture and other atmospheric influences are quite high [104]. The sulfur content in slag waste is less than 1.0%. Thus, the content of harmful impurities in the investigated ash-slag waste is insignificant and does not restrict their application.

In the ternary diagram compiled in [167] for the chemical composition of the ash from burning solid fuel of various deposits, the ash of Novocherkasskaya TPP falls into the region of acid ash (Fig. 4.11).

According to the basicity modulus, which takes into account the ratio of acid and basic oxides, the investigated ash-slag wastes are classified as acidic: for ash, the basicity modulus is 0.064, and for slag it is 0.086, i.e., less than one. According to the size of the clay-glandular module, taking into account the ratios of alumina and silica, ash-slag wastes are classified as active additives ($M_w = 0.50–0.69$) [77, 78]. However, the ashes themselves do not harden, but harden when mixed with water in the presence of lime or cement, i.e., exhibit the properties of pozzolans. This feature of ash is manifested during the structural formation (hardening) of the mass of the subgrade or the structural layer of pavement. In slag waste in trace amounts contains trace elements. Table 4.5 shows the results of a spectral analysis of ash-slag waste from Novocherkasskaya TPP.

According to researchers [75, 124], the following trace elements are contained in ash: boron, molybdenum, germanium, gallium, uranium, arsenic, vanadium, mercury, zinc, lead, cobalt, nickel, fluorine, and others.

4.3.4 Mineralogical Composition of Ash-slag Waste

The mineralogical and chemical compositions of ash and slag are determined mainly by the mineral content of the fuel and the changes that it undergoes during high-temperature processing in boilers of thermal power plants. In the furnaces of various designs, the behavior of the mineral part of coal is not the same even when burning the

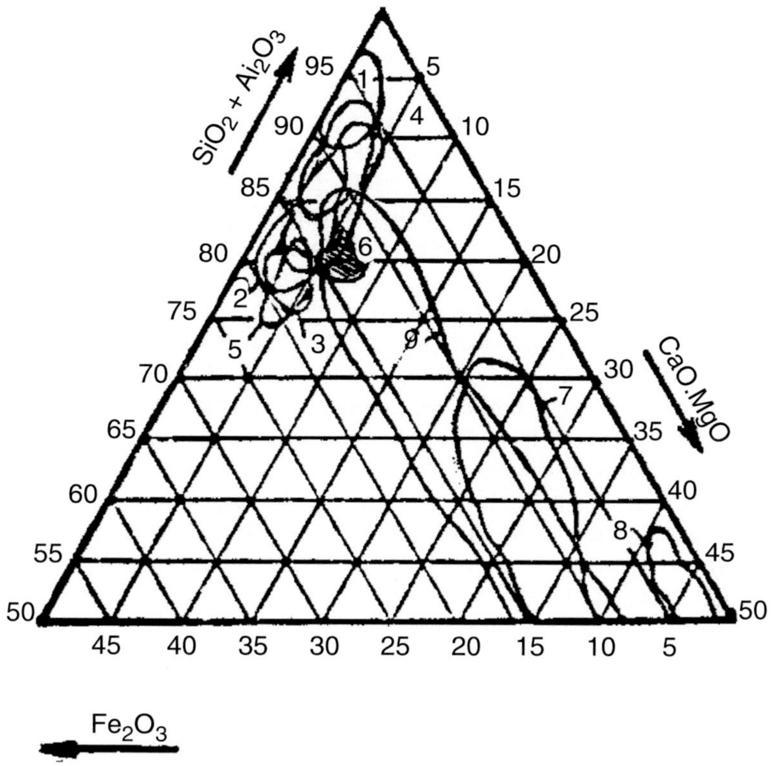

Fig. 4.11 Triple diagram of the chemical composition of solid fly ash of various deposits: Ekibastuz coal, Donetsk anthracite, Lviv and Volynsk, Kuznetsk, Donetsk gas coal, near Moscow, Kansk-Achinsk, Estonian slate, peat

Table 4.5 Data of spectral analysis of ash-slag waste, gram/ton of dry matter

Probe	Be	Co	Mn	Cr	Pb	V	Ni
Dry ash							
1	3.2	23.0	470.9	100.2	88.1	94.2	73.1
2	2.9	18.0	430.4	96.1	86.1	85.1	67.0
3	2.8	28.0	460.9	110.4	108.2	105.2	95.2
4	2.3	27.1	462.3	109.5	110.5	110.5	87.4
5	2.2	21.1	352.1	103.6	56.3	107.6	56.3
6	2.6	20.0	386.1	96.3	93.3	105.3	60.2
7	2.8	20.1	431.7	97.4	103.4	108.4	70.3
8	2.8	22.0	731.0	110.0	86.2	93.5	70.0
9	2.4	23.1	386.5	96.4	99.4	98.4	73.3
10	2.3	21.5	324.0	105.3	93.3	101.3	80.2
Hydraulic dump slag mixture							
11	1.7	25.0	315.6	121.2	24.0	106.2	95.2
12	2.1	23.4	324.0	120.3	30.1	99.3	84.2

same fuel. Therefore, the mineralogical composition of the ash will depend, in addition to the petrographic composition of the fuel, on the mode of its burning and calorific value. Most often, the mineral part of the fuel, in particular coal and anthracite coals, is represented by clays and minerals with the inclusion of dispersed sand fractions.

When burning solid fuel, its mineral part undergoes transformations associated with dehydration, dissociation, oxidation, polymorphic changes, interaction in the solid phase, and, in the melt, amorphization. Depending on the temperature and composition of the mineral part of the fuel, ash and slag can be formed: (i) without the appearance of a melt, (ii) with partial melting, and (iii) with complete melting of the initial mineral components. In the latter case, slags are obtained, which differ in the homogeneity of the phase composition.

The coarse material, in which quartz grains predominate, remains almost unchanged when burning solid fuel. Large (more than 0.05 mm) grains are dissected by cracks, sometimes melted. Most grains are coated with glassy melting products of clay. Feldspar grains melt significantly more, especially if their surface has been affected by weathering processes.

Iron sulfides (marcasite, pyrite) in the process of fuel combustion are oxidized to SO_2 and hematite or magnetite. Iron hydroxides at 250–450 °C dehydrate and pass into hematite. Carbonates in the range of 700–900 °C dissociate and pass into cryptocrystalline CaO and MgO.

The organic part of the initial fuel during the combustion process undergoes transformations associated with the removal of volatile substances at 200–400 °C and the transition of organics to semicoke (non-sintering particles) and coke (sintered particles). At subsequent stages, coke is oxidized to CO and CO_2. Due to the temperature nonuniformity in the furnace space, the completeness of these transformations varies significantly, and unburnt organic residues with various properties can be contained in the combustion products: from slightly changed, charred plant fibers to coke and soot carbon. Thus, inorganic and organic phases can be conventionally distinguished in the composition of ash and slag. The quantitative ratios of these phases depend on the type of fuel burned, the mode of combustion, and the method of removal of residual products. The inorganic phase includes amorphous and crystalline components. The amorphous component can be represented by glass and an amorphized clay substance. The crystalline component includes slightly modified grains of the minerals of the initial fuel (quartz, feldspars, and other thermally stable minerals), as well as crystalline neoplasms arising from the combustion of fuel (mullite, hematite, magnetite, calcium silicates, and others).

When burning fuel, there are a number of transformations that fail to complete before the onset of the equilibrium state. The product of this incomplete equilibrium is the vitreous phase. Glass in ashes can be silicate, aluminosilicate, and iron-aluminosilicate composition. Depending on the composition, the glass may be colorless, yellow, brown, or black.

The type and number of crystalline neoplasms in the ash can be determined by complex studies, including X-ray phase analysis and light microscopy. To determine the amorphous component of ash, quantitative mineralogical analysis in immersion preparations should be used.

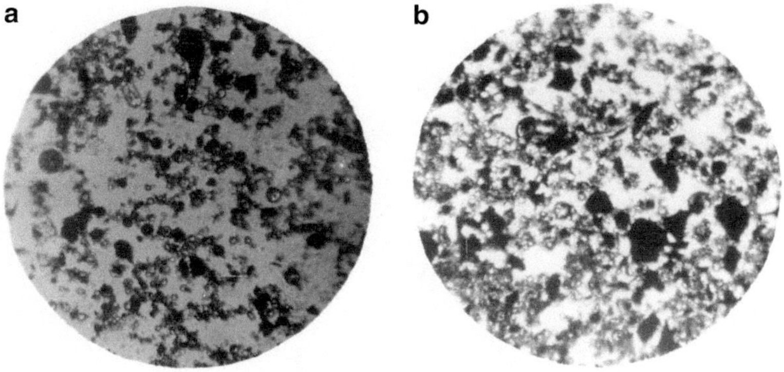

Fig. 4.12 Microphotography of fly ash in immersion (×0.63 Nicoli H)

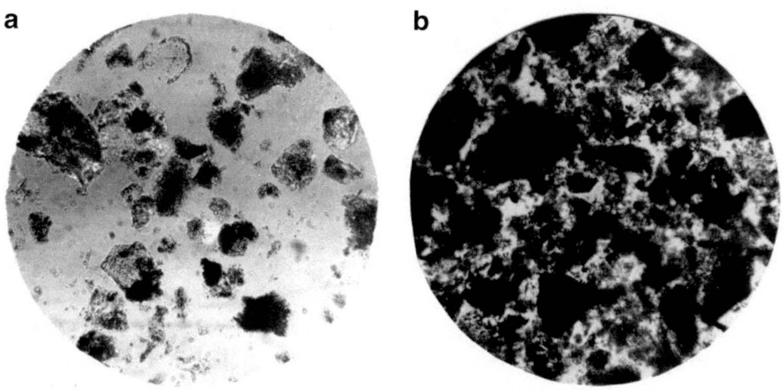

Fig. 4.13 Microphotography of immersion slag (×0.63 Nicoli H)

When studying the mineral composition of ash-slag waste of Novocherkasskaya TTP, quartz, magnetite, hematite, feldspars, amorphized clay material, mullite, and glass phase are determined. The mineral part consists mainly of clay minerals: hydromica mica and montmorillonites.

A microscopic (×63) examination of ash (see Fig. 4.12) and slag (see Fig. 4.13) reveals that their particles are different not only in shape, but also in size and composition. Particles of ash and slag are represented by quartz and glass grains of the most diverse shapes, from round to angular. Particle sizes: 0.02–0.04 mm (ash) and 0.25–0.50 mm (slag). The glass is transparent and painted in faint yellow, brown, and dark brown tones with refractive indices of 1.48, 1.55, and 1.67, characteristic of feldspar and clay glasses. Dark particles have an irregular angular shape and large sizes: 1–2 mm. In slag, they stick together in layers; quartz grains subjected to temperature processes have an iris.

The ash fraction with a particle size of less than 0.001 mm is represented by an amorphized clay substance with a refractive index of 1.57 close to the refractive

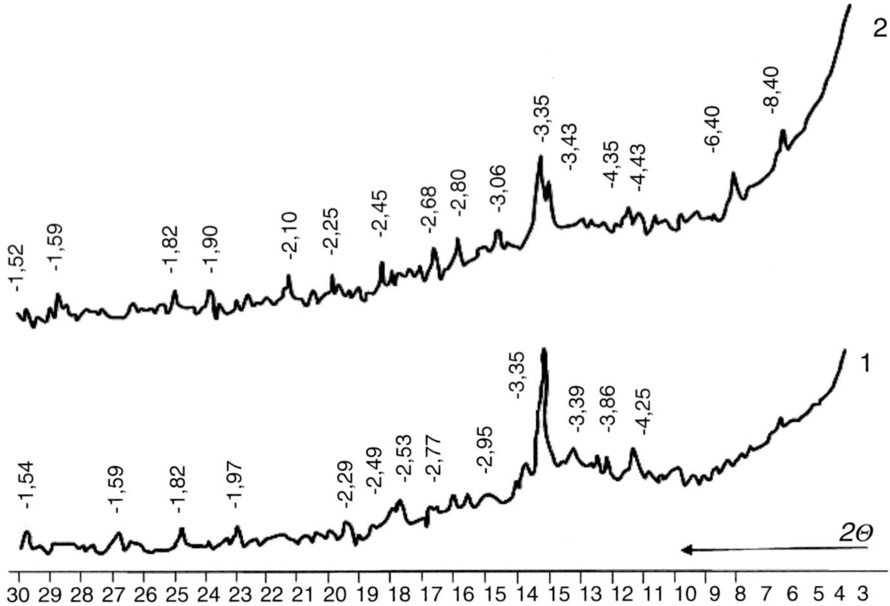

Fig. 4.14 Roentgenograms of ash-slag waste: 1—fly ash; 2—slag

index of kaolinite. In a fraction of less than 0.005 mm, colorless and colored spherical glassy particles, glass, fused grains of feldspar, quartz, and finely divided calcite crystals were found. A fraction of particles greater than 0.005 mm is predominantly represented by spherical glassy particles, in which, along with their colorless porous species, there are particles with a highly corroded surface coated with finely divided residual fuel, and spherical particles of complex composition. In spherical glassy particles, single needle-shaped mullite crystals, partially fused grains of quartz, and other minerals are observed.

Of the glasses, mainly glass of aluminosilicate composition predominates. X-ray structural and thermal analyzes of ash and slag (see Fig. 4.14) confirm that the crystalline phase is mainly represented by quartz, tridymite, aluminum, and calcium oxides; feldspars in the form of albite and galenite are present. On the thermograms (see Fig. 4.15), the endothermal effect at 150 °C, characteristic of the removal of free water, is traced. The endothermal effect at 460 °C is associated with the dehydration of clay minerals, the release of constitutional water, and the destruction of the crystal lattice. Exothermal effects at 130 and 280 °C corresponded to transition of amorphous iron to crystalline. Exothermal effects at 680 and 800 °C were also observed, accompanied by a polymorphic transformation of quartz, dissociation of carbonates, and burnout of carbonized carbon particles.

From the data of thermal analysis, it follows that in the temperature range of 20–1000 °C in the ash and slag approximately the same physicochemical transformations occur. However, endothermal and exothermal effects in ash have a greater

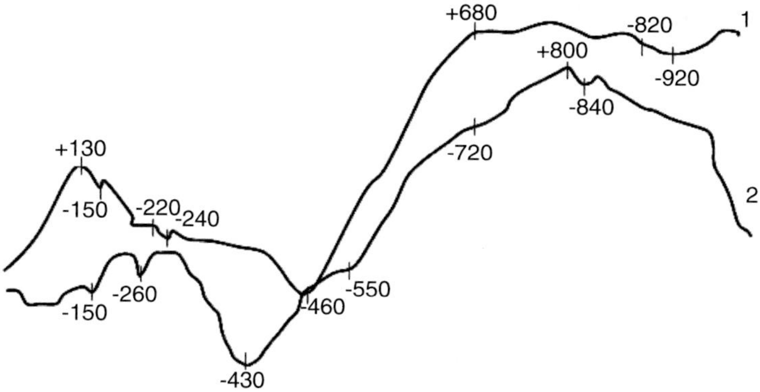

Fig. 4.15 Thermograms of ash-slag waste: 1—fly ash; 2—slag

peak depth and weight loss compared to slag (total weight loss is 7%). The slag mass loss curve is flatter, and the bulk of the substance (3.3%) is lost after 550 °C—the beginning of the burning of carbonaceous substances and the decomposition of carbonates.

4.4 Hydraulic Activity of Ash and Burnt Mine Dumps

Hydraulic activity is the most important property of ash-slag waste and burnt mine dumps. This feature allows the use of these wastes as active mineral additives. These materials do not have the ability to interact directly with water. However, in a mixture with binders (lime, cement) they exhibit pozzolanic activity, that is, the ability at ordinary temperatures to bind calcium hydroxide to form insoluble compounds. With the accumulation of these neoplasms, there is the possibility of hydraulic hardening of binders from mixtures of lime or Portland cement with pozzolanic additives. This is characteristic of volcanic rocks (pozzolans), which gave the name to this phenomenon.

Clay minerals predominate in the mineral part of solid fuels, as well as in the initial mine rocks. The most characteristic type of clay minerals in coal seams is kaolinite ($Al_2O_3 \cdot 2SiO_2 \cdot 2H_2O$). The occurrence of the activity of burnt rocks and ash-slag waste is based on a change in clay matter during physicochemical processes that occur during high-temperature firing of carbohydrate rocks and burning solid fuel. When firing in a dump or during combustion in the boiler furnaces, dehydration and amorphization of clay material and formation of the following phases occur [77, 157]:

(i) Metakaolinite is an incompletely dehydrated and amorphized clay substance that retains a distorted crystal lattice and is capable of hydration
(ii) Amorphized, but not sintered, characterized by a highly developed surface, consisting mainly of a thin mechanical mixture of amorphous silica and alumina

(iii) Sintered and partially vitrified substance having a significantly lower specific surface area

The highest specific surface and the highest hydraulic activity are characterized by the first phase. Having a large specific surface, metakaolinite actively reacts with calcium hydroxide at ordinary temperatures with the formation of calcium hydrosilicates and hydrogenite by the reaction

$$Al_2O_3 \bullet 2SiO_2 + 3Ca(OH)_2 + (n + 5)H_2O = 2CaO\bullet Al_2O_3 \bullet SiO_2 \bullet 8H_2O + CaO\bullet SiO_2 \bullet H_2O$$

With a decrease in the specific surface and partial vitrification of amorphous products of high-temperature processing of clay minerals, the hydraulic activity of the components of burnt rocks, ashes, and slags decreases.

Along with the clay substance, sandstones, which are cemented quartz grains, are also subjected to pyro-processes. In the process of firing, quartz, having undergone a series of transformations, returns to the low-temperature modification of β-quartz, but with a damaged structural lattice. Glandular impurities under the influence of firing processes also undergo changes and move into more active forms.

Thus, the main active components of burnt rocks, ash, and slag, capable of further interaction, are metakaolinite, SiO_2, Al_2O_3, and Fe_2O_3, in which the structure of the starting minerals is disturbed under the influence of pyro-processes. The presence of these components is associated with a violation of the crystal lattice of clay minerals during thermal treatments and the appearance of a certain energy potential in thermally modified products. This new state of matter also causes the ability of burnt rocks and ashes to hydrate and absorb lime from solution.

The existing differences in the material composition of the initial mine rocks and solid fuels, as well as in the conditions of their combustion, affect the degree of activation and reactivity of individual components of burnt rocks and ash-slag waste. Unlike ashes, burnt rocks contain a significant amount of clay, and glandular and siliceous hydraulic components. Burnt rocks exhibit the properties of active clay, and ashes demonstrate the properties of medium-active silicate-glandular additives. The most active component of ash and slag is vitrified substance and glass phase. The activity of the ashes is mainly due to the presence of a glass phase.

At elevated temperatures and water vapor pressure, some crystalline components of the waste are active (quartz, field wheat, calcium aluminosilicates). The nature of the interaction of the active components of the waste with aqueous solutions of calcium hydroxide depends on the specific surface area of these components and their solubility. Having a large specific surface, metakaolinite actively reacts with calcium hydroxide at ordinary temperatures with the formation of crystalline and gel-like calcium hydrosilicates of the CCH group (1).

The activity of amorphous $A1_2O_3$ and SiO_2 formed at high temperatures is noticeably lower, which is explained by a decrease in the specific surface and solubility due to sintering and crystallization of neoplasms (mullite, cristobalite). High-temperature forms of silica and heat-treated quartz, when interacting with calcium hydroxide, form hydrosilicates of the CCH group (1). The initial quartz, which has not undergone firing, is practically inert under these conditions.

High-temperature sintering and melting of clay minerals sharply reduce their specific surface area and, correspondingly, activity; therefore, the glass phase of ash-slag waste is inactive at ordinary temperatures. The use of thermal water treatment sharply increases the activity of products of amorphization and melting of clay material. At the same time, the rate and volume of absorption of calcium oxide from the solution by these materials increase. In this case, the activity of the sintered and vitrified amorphous phases with a small specific surface increases to the greatest extent. The greatest contribution to the activity is made by the glass phase. The influence of ferrous oxides should be taken into account when assessing the activity of burnt rocks.

When using ash-slag waste and burnt rocks as an active mineral additive, it becomes necessary to evaluate their pozzolanic activity. Methods for assessing the activity of additives exhibiting the properties of pozzolans are diverse:

(i) At the end of the setting of a test of normal density, consisting of supplements, hydrated lime, and gypsum: the end of setting is not later than 7 days after mixing.
(ii) On water resistance of the seized test: within 3 days of being in water in samples containing the additive, erosion is not detected.
(iii) In terms of compressive strength of steamed samples from dough containing an additive: compressive strength should be at least 3 MPa.
(iv) By its ability to absorb dye from solution.

The activity of mineral additives can be most objectively assessed by the degree of their involvement in the pozzolanic reaction with calcium hydroxide released during the hydrolysis of Portland cement silicates.

Figure 4.16 presents a comparative assessment of the involvement of ash and burnt rock in the reaction with calcium hydroxide. This technique allows a comparative assessment of the activity of pozzolanic additives. Activity is evaluated by the degree of involvement of the additive in the reaction (ρ) and is expressed by the ratio of the mass of the additive that has entered into the reaction to the mass of the entire added additive. From the data obtained it follows that the pozzolanic reaction begins at a fairly early age, but a noticeable amount of the additive is involved in the reaction by 7 days. Subsequently, a slow increase in the amount of reacted ash or burnt rock occurs. The data obtained are consistent with the dispersion of the additives and the features of their amorphous phases, showing pozzolanic activity. From the above data it follows that in the process of hydration, finely ground burnt rock is more actively involved (curve 1). The burnt rock consists mainly of amorphized clay material, and this is the main reason for its active involvement in the pozzolanic reaction. The effect of the pozzolanic effect increases with increasing specific surface area of the additive (curves 2 and 3). This is explained by an increase in physicochemical activity along with chemical activity due to the extremely developed surface of the particles.

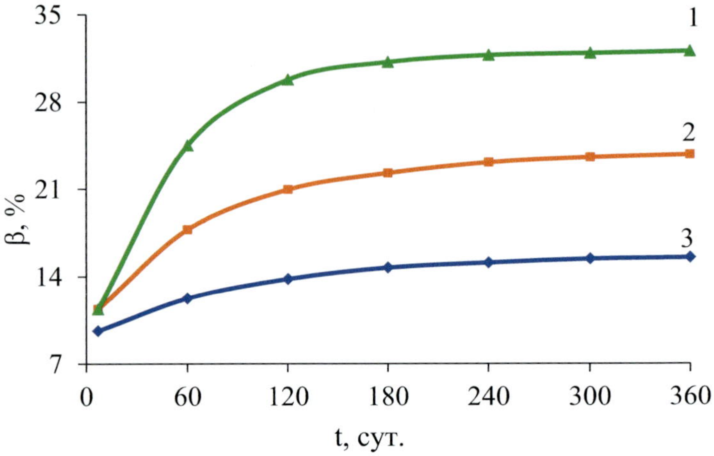

Fig. 4.16 Kinetics of the involvement of additives in the reaction with calcium hydroxide: 1—burnt rock, specific surface is 600 m^2/kg; 2—fly ash, specific surface area is 350 m^2/kg; 3—slag, specific surface is 200 m^2/kg

According to the photocolorimetric method, which allows estimating the activity of ash-slag waste and burnt rocks by their ability to absorb dye from solution, they are classified as adsorbents with medium activity. Based on the clay-glandular module (0.5–0.7), which takes into account the chemical and mineralogical composition, the studied waste also belongs to the group of additives with medium activity. According to the microcalorimetric evaluation of the activity of the additives, developed by G.I. Knigina et al. [77, 78], ash-slag wastes by their activity belong to the middle group (II group of activity), burnt rocks are related to mudstones, and siltstones to active and moderately active (II group of activity).

The data obtained are consistent with the dispersion of the additives and the features of their amorphous and vitreous phases exhibiting pozzolanic activity. According to the amount of fine-ground burnt rock that has reacted, we can conclude that its participation in the hydration process is more intense. The burnt rock consists mainly of amorphized clay material; this is the reason for its active involvement in the pozzolanic reaction. The effect of the pozzolanic effect increases with increasing specific surface area of the additive. Coarse ash is less involved in the reaction. It should be noted that a more complete involvement of the additive in the reaction causes a more intense binding of calcium hydroxide, which in turn enhances the hydrolysis of cement silicates and leads to an additional increase in the degree of hydration. This affects the hardening kinetics, structure, and properties of the hardened material. Thus, an indicator of the degree of involvement of the mineral additive in the reaction can be used for a comparative assessment of the pozzolanic activity of ash and burnt rocks.

The simplest method for assessing the quality of various types of mineral additives is the ability of samples of binders containing additives and having a strength of 7 days to withstand water resistance tests after 3 days of water saturation. The

Table 4.6 Assessment of the quality of additives from technogenic raw materials

	Structure, weight %		Compressive strength at age of 7 days, MPa		Condition of samples after 3 days of water saturation	
No.	Fly ash	Burnt rock	Fly ash	Burnt rock	Fly ash	Burnt rock
1	70	30	8.6	7.4	No erosion of the samples was detected; the sharpness of their faces is preserved	
2	60	40	7,5	6.8		
3	50	50	6.3	5.1		
4	40	60	5.2	4.8		
5	30	70	4.1	3.4		

results of assessing the quality of additives from burnt rock and fly ash at various proportions with cement are given in Table 4.6.

It is possible to increase the specific surface area, activity, and adsorption capacity of burnt rock, ash, slag, or a mixture thereof with joint grinding with activator additives. When grinding, the effect of mechanical and chemical activation of substances is manifested, its reactivity increases. When grinding any substance, the energy spent on its grinding is spent on breaking bonds in the crystal and creating new surfaces [13, 85]. When grinding ash, slag waste, and burnt rock, the vitreous shell around the grains of ash, slag, and rock is destroyed. The breaking of structural bonds is accompanied by the appearance on the exposed surface of particles of ash and burnt rock of active centers of various nature and strength. The resulting active centers, depending on the conditions, can participate in various adsorption processes and solid-phase reactions.

Figure 4.17 shows the change in the specific surface of a mixture of ash and burnt rock, in a ratio of 1:1, when combined with an alkali-containing additive. As such an additive, solid waste from the spent solution of a caustic soda electrolysis cell was used. It was experimentally proved that the specific surface area of hydraulic additives from ash-slag waste and burnt mine rocks should be at least 500 m^2/kg. This conclusion is consistent with the author's statement [154] that the optimum dispersion of cement additives should be 120–200 m^2/kg higher than cement dispersion.

The activity of mineral additives was evaluated by adsorption of the dye, methylene blue [76]. A characteristic of adsorption activity is a change in the optical density of the dye (initial concentration) before dye treatment and after treatment (final dye concentration):

$$\Delta D = D_1 - \mathrm{D}_2,$$

where D_1 and D_2 are the dye densities of the initial concentration, before and after processing a sample of mineral additives, respectively.

Figure 4.18 shows the change in the adsorption activity of a mineral additive containing fly ash and burnt rock in a ratio of 1:1 of the content of alkali-containing additives. From the given Figs. 4.17 and 4.18, it follows that when co-grinding the

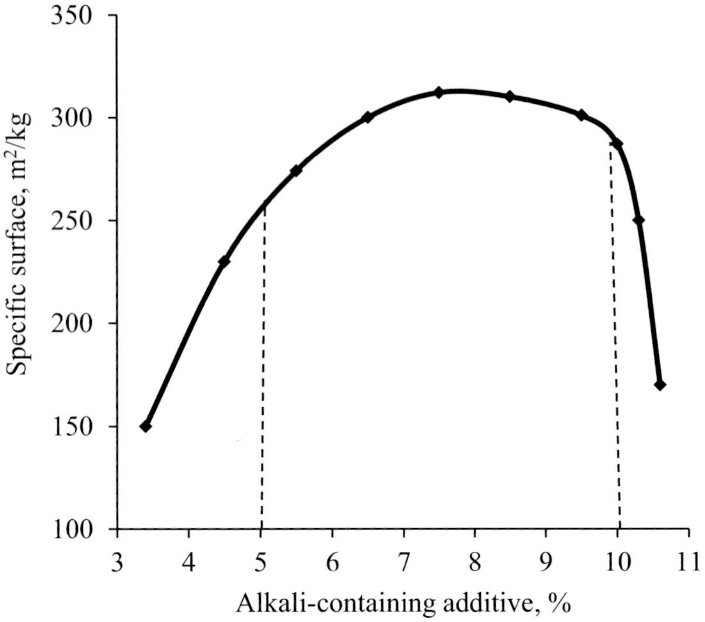

Fig. 4.17 Change in the specific surface of the additive composition (fly ash:burnt rock = 1:1)

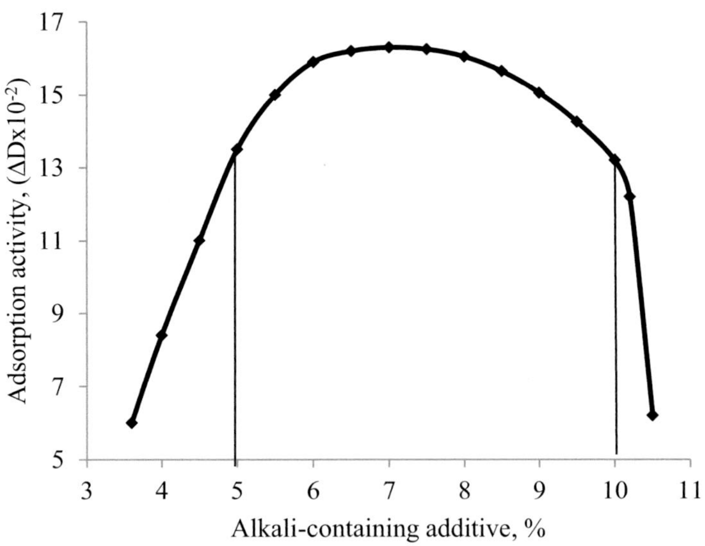

Fig. 4.18 Change in the adsorption activity of a mixture of fly ash and burnt rock in a ratio of 1:1 with joint grinding

Table 4.7 Adsorption activity of mineral additives (ΔD)

Physical condition of the mineral supplement	Evaluation of adsorption activity by ΔD		
	Fly ash	Burnt rock	$\frac{\text{Fly ash}}{\text{Burnt rock}} = 1:1$
Initial	0.031	0.022	0.05
After grinding	0.071	0.062	0.09
After grinding with an alkaline additive	0.13	0.11	0.16
Increased adsorption activity (times)	4.19	5.0	3.2

optimal values of the specific surface and adsorption activity of the mineral additive account for the content of alkaline additives from 5 to 10%.

The performed comparative determinations of the adsorption capacity of the initial fly ash and burnt ash and their mixture before and after co-milling with an alkaline additive confirm an increase in adsorption activity by more than 3.2 times (see Table 4.7). From the above data it follows that fly ash is more active in comparison with burnt rock. However, the activation ability of burnt rock is more pronounced than that of fly ash. The greatest activation effect was observed in the mixture of fly ash and burnt rock. Obviously, there is a synergistic effect, which is confirmed by increased values of ΔD for a mixture of ash and rock.

The established mechanism of activation of fly ash and burnt ash was the basis for the development of a method for producing a binder for concrete and mortar [24].

The main criterion in assessing the activity of ash or finely divided burnt rock of mine dumps is their ability to interact with the hydration products of clinker minerals without reducing the strength of the hardening stone, or to absorb lime from a lime mortar. An indicator of activity is the proportion of pulverized ash or finely divided burnt rock in the composition of a complex binder, in which there is no decrease in the strength of cement samples with ash or finely divided burnt rock in comparison with conventional, without additives.

The use of such additives allows one to save the consumption of clinker minerals while improving the construction and technical properties and operational characteristics of artificial composites. The research results are confirmed by the practice of using ash and slag waste and burnt rocks of mine dumps.

Chapter 5
The Use of Burnt Rocks of Mine Dumps and Ash-slag Waste in Road Construction

Abstract This chapter discusses the use of the burnt rocks of mine dumps and ash-slag waste for the building of subgrade road foundations, and as fillers of these products in the constructional layers of pavements on the base of practical experience accumulated in the Rostov region (Russia). For each constituent of road construction, certain materials are treated, which can allow attainment of the optimal properties and long-life service of road pavement. In particular, hardening kinetics of compositions containing technogenic raw materials are studied and also application of these materials in asphalt concrete and concrete coatings. The main characteristics, defining strength and durability of the constructions, are considered as follows: water saturation, water resistance, frost resistance, adhesion of bitumen to the mineral, abrasion, cement saving, etc. Moreover, replacement of ordinary concrete by the concrete products based on mineral fillers and binders is discussed. Compression strength, bending strength, and kinetics of the development of linear strains of concrete during cyclic tests in the conditions of aggressive medium are studied.

Keywords Subgrade composition · Inorganic and organic binders · Mine rock characterization · Slag mixture · Pavement base layers · Clay and silt particles · Hardening kinetics · Compressive strength · Frost resistance

5.1 Subgrade Composition

In road construction, burnt rocks of mine dumps and ash-slag mixtures can be used for the construction of subgrade road foundations, and as fillers of these products in the constructional layers of pavements. Low-active, fine-disperse dry ash and burnt mine rock are used as a mineral additive to inorganic and organic binders.

This is especially true if the object of consumption of waste is in close proximity to the location of technogenic raw materials. When using such raw materials, in addition to saving natural mineral raw materials, the cost of transportation for the

A. A. Lyapin et al., *Improving Road Pavement Characteristics*, Innovation and Discovery in Russian Science and Engineering,
https://doi.org/10.1007/978-3-030-59230-1_5

delivery of raw materials to the place of consumption is reduced. Some environmental problems are being solved to reduce the negative impact of the waste dump on the environment near its location.

An example of such an integrated approach to the problems of energy, resource saving, and recycling of industrial waste is the use of mine dump rocks for the construction of subgrade during the construction of an access railway to the public railway in Belokalitvinsky district of the Rostov region (Russia) [27, 28, 33]. The length of the access road connecting the Sadkinskaya mine with a common highway is equal to 27 km.

Studies on the assessment of the possibility of using mine rocks for the construction of the subgrade of the railway track are carried out on the example of waste dumps of the Sevryugovskaya mine (dump No. 7) and Sinegorskaya mine (dump No. 17) located in the village Sinegorsky, Rostov region (Russia). The rocks of the studied mine dumps are classified as polymineral raw materials and are represented by clay (mudstone-siltstone) rocks, sand-clay, and slightly sandy; carbonaceous schists and sandstones are also found. Mudstone (from 50 to 60%), siltstones (from 10 to 35%), and sandstones (up to 10%) prevail in the composition of the dump; siliceous rocks and coal-mineral splices are found. According to the lithological composition, the rock mass of the dumps is sandy-clay. Clay rocks of carbon-bearing deposits are subject to compaction, as a result of which they turn into stonelike rocks. According to the grain composition in the mine dumps, fragments from 6 to 100 mm in size prevail. The rocks discharged from the mines have a grain size of up to 250 mm, and individual pieces reach 350–450 mm. Pieces larger than 600 mm are available, too.

Mine rocks, unlike conventional soils traditionally used for the construction of subgrade, have specific properties, and they should be attributed to special soils. The rocks of these dumps belong to the class of coarse-grained soils, which are non-cemented deposits of fragments between which there are no structural bonds (pieces of rock, crashed stones, pebble, wood, gravel, sand) thermally changed in the dump under conditions of prolonged self-firing at a temperature of about 1000 °C. The composition of rocks is given in Table 5.1.

According to the content of calcium oxides, the studied rocks belong to low calcium (CaO <20%) and by the sum of silicon and aluminum oxides they belong to siliceous-alumina. The content of carbonaceous impurities does not exceed 7.0%. The types of carbonaceous impurities found in dumps are resistant to oxidation, and their durability when exposed to moisture is quite high. There are no restrictions on the specific effective activity of natural radionuclides on the use of rocks of the studied dumps (A_{ef} <370 Bq/kg).

Table 5.1 Chemical composition of the rocks of mine dumps

Mine	SiO_2	Al_2O_3	Fe_2O_3	CaO	MgO	TiO_2	K_2O	Na_2O	SO_3 Com.	IL
Sinegorskaya	55.73	18.68	7.50	1.89	2.23	0.95	2.95	2.30	2.77	4.50
Sevryugovskaya	57.35	19.95	5.47	1.43	1.95	0.99	3.50	1.09	1.62	6.16

IL ignition loss

Testing of the physical and mechanical properties of mine rock as soil for subgrade was carried out in accordance with applicable regulations for soil of subgrade, as well as technical specifications for materials from burnt rocks of mine heaps, which set out requirements that take into account the specific properties of the rocks. The results of studies of the physico-mechanical properties of materials from mine rocks of the studied dumps are given in Tables 5.2, 5.3, and 5.4. The curve of standard rock mass compaction is shown in Fig. 5.1.

The obtained quality indicators of mine rocks of dumps correspond to the requirements for soils used for the construction of the subgrade. In terms of physical and mechanical properties and mineral composition, the rocks of the studied dumps are heterogeneous. This must be taken into account when preparing rocks, aimed at averaging technogenic raw materials by particle size distribution, and physical and mechanical properties. To construct the subgrade of the access railway track, the mine rocks of the dumps were averaged by multiple transshipment with a bulldozer during shipment from the dump. The maximum size of pieces of crushed rock should be no more than 70 mm.

The structural design of the subgrade of the access railway track is shown in Fig. 5.2.

The technology of building subgrade from mine rocks is similar to layer-by-layer building of the embankment from the traditional soil used. The required density of mine dump rocks in the body of the embankment should be applied according to the design documentation. Optimum humidity and maximum density should be determined using the standard compaction method.

The rock is transported to the site of construction of the track. The building process consists of sequential layering of the rock mass. The largest allowable size of pieces of mine rock from the dump should not exceed 70 mm. The sprinkled layer

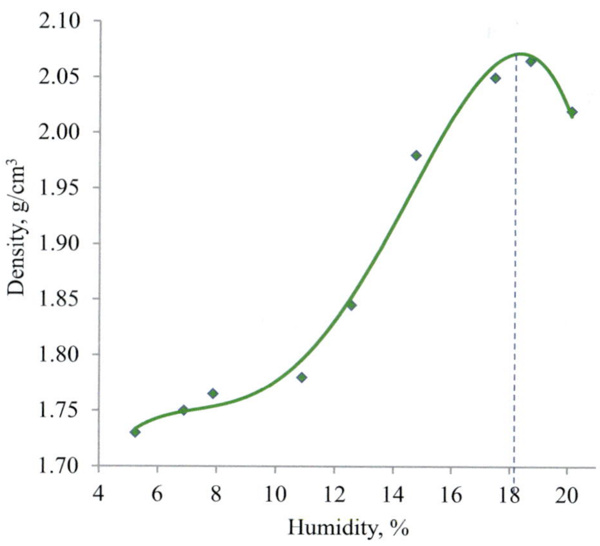

Fig. 5.1 Standard seal curve: the maximum density of the rock mass is 2.06 g/cm^3; optimum humidity is 18.2%

Table 5.2 Mine rock mass characterization

Key quality indicators	Values of indicators	
Specific weight, g/cm^3	2.57	
Volumetric weight, kg/m^3	1450	
Volumetric weight of the skeleton, g/cm^3	1.57	
Grade of crushed stone fraction in strength at crushability test: Dry state		
From >5 to 10 mm	300	
From 10 to 20 mm	400	
From 20 to 40 mm	600	
Water resistance grade	W1	
Grade of crushed stone fraction on frost resistance		
From >5 to 10 mm	F15	
From 10 to 20 mm	F25	
From 20 to 40 mm	F25	
Content of grains of weak rocks in crushed stone, %	9.8	
Sieve analysis, residues on sieves, % by weight	Individual	Total
40 mm	21.28	21.28
20 mm	23.49	44.77
10 mm	17.21	61.98
5 mm	11.63	73.61
2.5 mm	11.20	84.81
1.25 mm	2.71	87.52
0.63 mm	4.27	91.79
0.315 mm	5.63	97.42
0.16 mm	2.31	99.73
<0.16 mm	0.27	–
Resistance of crushed stone fraction against all types of decay, mass loss during decay, %	1.78	
Water absorption, %	2.13	
Humidity at the yield strength, %	28.3	
Humidity at the rolling edge, %	26.0	
Plasticity	2.3	
Group on degree of bulging	I (non-bulged)	
Filtration coefficient, m/day	5.88	
Porosity	0.81	
Compaction ratio	0.98	
Density of dry soil with standard compaction, g/cm^3	2.06	
Optimum humidity with standard compaction, %	18.8	
Angle of internal friction, degrees	38	
Adhesion, MPa	0.02	

Table 5.3 Characteristics of crushed stone from mine rocks

Main indicators of the quality of crushed stone	Values	
	First crushing	Second crushing
Crush strength grade for fractions:		
From >5 to 10 mm	400	800
From 10 to 20 mm	800	1200
From 20 to 40 mm	800	1000
Frost resistance grade for fractions:		
From >5 to 10 mm	F15	F25
From 10 to 20 mm	F25	F50
From 20 to 40 mm	F25	F50
Abrasion grade	A1	A1
Water resistance grade	W1	W1
Bulk density, kg/m^3 for fractions:		
From >5 to 10 mm	1150	1080
From 10 to 20 mm	1130	1050
From 20 to 40 mm	1110	1040
True density, g/cm^3	2.56–2.63	
Content of dust and clay particles, %	1.07–1.35	
Content of lamellar and needle-shaped grains, %	6.15–21.4	
Content of grains of weak rocks, %	3.9–8.6	
Clay content in lumps, %	Absent	
Resistance to all types of decay, weight loss during decay, %, for fractions:		
From >5 to 10 mm	0.68	0.58
From 10 to 20 mm	1.02	0.47
From 20 to 40 mm	0.77	0.64
Water absorption, %	1.13–0.85	

Table 5.4 Characterization of screenings of crushing from mine rocks

Main indicators of the quality of crushed stone	Values	
	First crushing	Second crushing
Fineness modulus	3.17	3.88
Bulk density, kg/m^3	1310	1240
True density, g/cm^3	2.58	
Content of dust and clay particles, %	2.34	1.58
Clay content in lumps, %	Absent	
Coefficient of filtration, m/day	5.75	7.30
Gravel content, %	22–32	
Grade on strength of crushed stone fraction	400; 600	

is leveled by a bulldozer. The rock is moistened to optimum moisture. After planning the surface, the rock is compacted with middle or heavy rollers with metal rollers. Dump mine rocks are well compacted. When compacting such rocks, it is most effective to use vibratory rollers in combination with heavy rollers on pneumatic

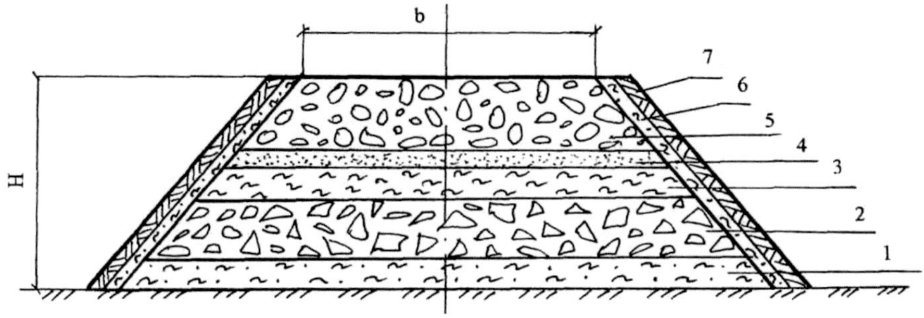

Fig. 5.2 Subgrade scheme: 1—underlying layer of loam; 2—fractionated burnt mine rock; 3—insulating layer of clay; 4—sand; 5—crushed stone from natural stone; 6—insulating layer of loam; 7—fertile soil; *b* is the width of the main site; *H* is the embankment height

tires with 2–4 passes along one track. When compacted by rollers, part of the rock is destroyed with the formation of fine earth (fractions less than 2 mm). This plays a positive role in ensuring close packing of the rock in the body of the embankment. Moreover, burnt mine rocks have one feature. The finely dispersed fraction of burnt and burnt-out rocks has pozzolanic properties; that is, it is capable of exhibiting latent physicochemical and hydraulic activity. This feature of the rocks allows us to attribute them to a self-sealing system. Due to this, over time, self-hardening of the rock mass occurs, which creates additional strength and stability of the embankment from the mine rocks.

The ash content of the rock mass is high (99.0%). However, due to the fact that dumps contain impurities of non-burnt rocks and it is not always possible to achieve proper averaging of the rock, the option of laying waste rock with a protective and insulating screen made of clay soil or loam is not possible. Moreover, it is necessary to reinforce the shoulders in accordance with the requirements of building codes, i.e., strengthening roadsides from cohesive soils. When planning slopes, apply a layer of fertile soil and sow perennial grasses on the slopes.

During operation of the railway track section, deviations in the geometric dimensions of the subgrade and embankment settlement exceeding the permissible limits were not found. Positive practical experience in operating the access road confirms the practical feasibility of using unconventional raw materials for the construction of the subgrade. Taking into account the low cost of materials from burnt rocks, the total cost of works on the construction of the subgrade is reduced. At the same time, mine rocks were utilized, two mine dumps were eliminated, and land suitable for further use was released.

During the construction of roads, according to the recommendations of design and regulatory documents, it is necessary to maximize the use of waste from thermal power plants suitable for use: fly ash, slag, and ash-slag mixtures. For the construction of the embankment of the subgrade, ash-slag mixtures of Novocherkasskaya TPP can be used. These wastes are classified as medium-active, have frost resistance (grade on frost resistance of the slag component of at least F15), and are nonporous.

Table 5.5 Characterization of slag mixture from dumps of Novocherkasskaya TPP

Main indicators of the quality	Values of indicators	
Specific weight, g/cm^3	2.45	
Volumetric weight, kg/m^3	1390	
Volumetric weight of the skeleton, g/cm^3	1.72	
Grade of crushed stone fraction in strength at crushability test: Dry state		
From >5 to 10 mm	300	
Water resistance grade	W1	
Grade of crushed stone fraction on frost resistance		
From >5 to 10 mm	No less F25	
Sieve analysis, residues on sieves, % by weight	Individual	Total
10 mm	11.52	11.52
5 mm	13.43	24.95
2.5 mm	15.14	40.09
1.25 mm	10.13	50.22
0.63 mm	12.27	62.49
0.315 mm	8.62	71.11
0.16 mm	13.65	84.76
< 0.16 mm	15.24	–
Resistance of crushed stone fraction against all types of decay, mass loss during decay, %	0.78	
Humidity at the yield strength, %	26.8	
Humidity at the rolling edge, %	25.1	
Plasticity	1.7	
Group on degree of bulging	I (non-bulged)	
Filtration coefficient, m/day	3.75	
Porosity	41.0	
Compaction ratio	0.98	
Density of dry soil with standard compaction, g/cm^3	2.13	
Optimum humidity with standard compaction, %	18.8	
Angle of internal friction, degrees	24°54′	
Adhesion, MPa	0.008	

The value of the relative deformation of frost heaving does not exceed 0.035. The main indicators of the properties of slag mixtures from dumps of Novocherkasskaya TPP are given in Table. 5.5.

Based on the value of relative frost bulging strain less than 0.035, ash-slag mixtures according to [68] are used to construct embankments of the subgrade without restrictions.

5.2 Materials from Technogenic Raw Materials for Pavement Base Layers

Road bases are the main bearing layers of pavement on which the road surfaces are located. The purpose of the base is the withstanding of the load from cars and through the coating and its distribution on the subgrade. In the road industry, an urgent area of research is the development of road-building materials, the physical and mechanical properties and technical characteristics of which would provide increased durability of pavement structures. Special requirements are applied to the materials used for the construction of substrates. The material of the base layer must be able to self-strengthen over time in order to provide resistance to the increasing load, since continuously increasing loads act on the bases during operation. Such a structural layer can be created using industrial waste.

In general, road bases consist of three layers. The lowest third layer is located directly on the subgrade and is called the underlying. The underlying layer, made primarily of local materials or industrial waste, should have a certain strength, characterized by a deformation modulus, and shear resistance, and should withstand frost heaving. The underlying layer, as a drainage layer, must have a filtering ability. As industrial waste for the construction of the underlying layers, burnt rocks of mine dumps and ash-slag mixtures can be used. The indicators of their properties comply with the requirements of regulatory documents [68] and provide reliable construction of additional layers in the construction of pavement. The studied burnt rocks of mine dumps and ash-slag waste are most suitable for use in road pavements. The specificity of the formation of burnt rocks and ash-slag waste contributes to the formation of material grains with a macro- and microporous structure characteristic of heat-insulating materials. From this it follows that the base-bearing layer of burnt rocks and ash-slag waste will also have heat-insulating properties. These materials can be used to create additional frost protection and thermal insulation layers. The particles of burnt rocks of mine dumps and slag mixtures have a good filtering ability. The underlying layer of these materials may also be draining. The presence of the underlying layer allows one to reduce the thickness of the overlying layer of expensive materials.

As an example, we give the results of using the layers of road bases of fly ash of Novocherkasskaya TPP, burnt rock of the Petrovsky mine dump, and dump slags of the Krasnosulinsky metallurgical plant. The waste is concentrated in the Rostov region (Russia). The use of ash-slag waste and metallurgical slag in the construction of road bases is more common than burnt rocks. As natural raw materials as the main fillers, screening crushing rocks of the Zhirnovsky quarry were used. The chemical composition of the raw materials used is given in Tables 5.6 and 5.7. The grain composition of the screening crushing is given in Table 5.8. Grade of crushed stone fraction of screening by crushing ability is 800, bulk density 1440 kg/m^3, and content of clay and silt particles 13.5%.

The chemical composition of the waste is classified as semi-acid raw materials, with a high content of coloring oxides. Screenings of crushed stone crushing by material composition are limestones, and calcium oxide prevails in the chemical composition.

Table 5.6 Chemical composition of mineral raw materials

Material	Content of oxides, weight, %									
	SiO_2	Al_2O_3	Fe_2O_3	CaO	MgO	SO_3	Na_2O	K_2O	P_2O_5	Ignition loss
Fly ash	54.92	21.10	6.13	4.96	1.86	0.52	0.97	1.13	0.93	7.25
Burnt mine rock	55.35	22.37	7.04	2.45	1.46	1.84	1.98	2.27	0.35	4.80
Screening crushing of crushed stone	0.78	1.13	0.52	52.2	0.47	0.37	0.13	0.18	0.04	44.5

Table 5.7 Chemical composition of metallurgical steel slag

Chemical composition, weight, %							
CaO	SiO_2	Al_2O_3	MgO	MnO	Fe	Fe_2O_3	S
32.0–45.0	29.0–37.0	3.0–6.0	3.0–8.0	1.8–3.5	8.0–23.0	1.0–3.5	0.7–2.0

Table 5.8 Grain composition of the screenings of crushed stone crushing

Diameter of holes of sieves, mm	10	5	2.5	1.25	0.63	0.315	0.16	<0.16
Total remains on sieves, % by weight	13.7	20.8	38.3	46.9	58.2	72.4	89.0	100

The chemical composition of steel slag is quite complex. Oxides Cr_2O_3, V_2O_3, TiO_2, WO_3, etc. may also be present in slags. Slags that have been in the dump for a long time are resistant to all types of decay.

The mineral part of the burnt rock and ash is mainly represented by clay material modified during the firing process, and the organic part is represented by modified coal modifications. Carbon impurities are present in the form of graphitized components, close to graphite, partially turning into semicoke varieties.

The main groups of substances are present in ash and rock: crystalline, amorphous, and organic. The amorphous component of the ash is represented by a glass phase and an amorphized clay substance. Crystalline minerals were found in the mineral composition of burnt rocks and ashes (Figs. 5.3 and 5.4):

(i) β-SiO_2 with $d = 3.35$; 2.29; 2.12; 1.82; 1.67 Å
(ii) β-Cristobalite with $d = 2.81$; 2.53; 2.49; 1.59; 1.54 Å
(iii) γ-Tridymite with $d = 4.29$; 3.21; 2.81; 2.49 Å
(iv) γ-Al_2O_3 with $d = 2.70$; 2.45; 2.29; 1.97 Å
(v) Mullite with $d = 2.20$; 1.69; 1.59 Å

In this series of experiments, fly ash, ground metallurgical slag, and burnt rock of the mine dump were used as additives to improve the particle size distribution and cementation conditions of screenings for crushing natural limestone. When selecting compositions for road surfaces, dry ash with a specific surface area of 350 m^2/kg was used. The burnt rock and slag were crushed to the same fineness as ash. Moreover, mixtures (binders) were prepared containing ash or burnt rock with cement. Dosed ash or ground rock and cement in 1:1 ratios were mixed and crushed together in a ball mill to a specific surface of 500–700 m^2/kg. Gypsum was introduced into the mixture (2–5% over 100%). With joint grinding, the vitreous shell around the grains

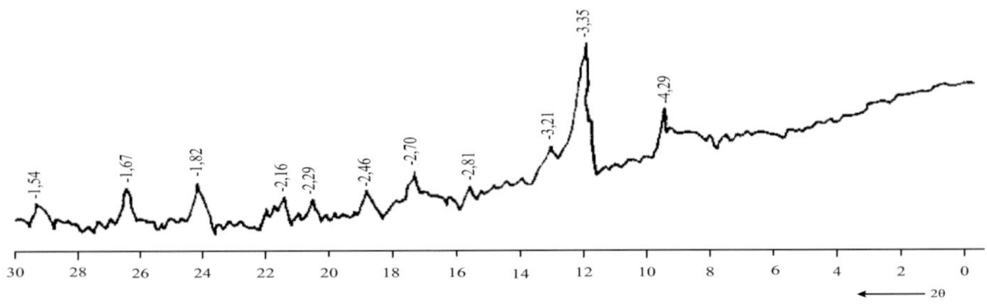

Fig. 5.3 Roentgenogram of burnt rock

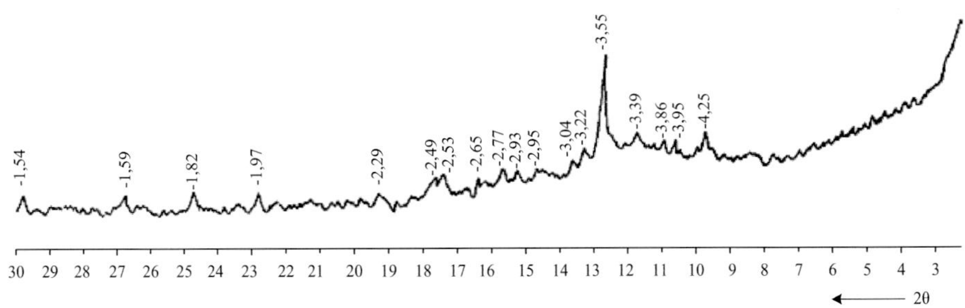

Fig. 5.4 Roentgenogram of fly ash

of ash and burnt rock is destroyed, structural defects appear, and active centers are exposed on the grain surface [38, 97]. All these processes contribute to an increase in the activity of the binder. Such a binder was used to strengthen screenings for crushing limestone. It should also be noted that when grinding waste products, their averaging occurs, since industrial waste stored in dumps is characterized by instability of their properties and structure.

The composition of the foundations of roads is based on the principle of contact hardening, which consists of the ability of dispersed systems of the optimal ratio of components and grain composition to form a solid monolith at the moment of convergence of particles during compaction. In accordance with the experimental design, mixtures were prepared for the formation of samples. Testing of the physico-mechanical properties of the designed compositions (see Table 5.9) was carried out on cylindrical samples with a height and a diameter of 50 mm. The components were mixed in a dry state and then with water to the optimum humidity (state for semidry pressing). The pressure of sample pressing was 10 MPa.

The strength of the samples was formed within 90 days under conditions of normal hardening with a temperature of (20 ± 2 °C) and relative humidity of air (95 ± 5%). The samples were tested for compressive strength at different times of hardening.

Figure 5.5 shows the results of the hardening kinetics of compositions containing various types of active additives, providing a set of their strength. The composition

Table 5.9 Compositions and their strength for road bases

Состав смеси, %							R_{com}, MPa			
				Low-quality binder			Days			
Cement	Fly ash	Ground burnt rock	Slag	C:FA	C:GBR	Screening crushing	28	56	90	R_b, MPa (28 days)
6.0	–	–	–	–	–	94.0	6.95	8.57	10.54	1.43
4.0	2.0	–	–	–	–	94.0	4.83	7.90	13.40	0.98
4.0		2.0	–	–	–	94.0	4.94	6.38	10.02	1.00
4.0	–	–	2.0	–	–	94.0	5.08	5.68	6.72	1.05
–	–	–	–	–	6.0	94.0	4.60	5.05	5.58	0.95
–	–	–	–	6.0	–	94.0	5.30	5.68	6.10	1.10

Note: *C:FA* cement/fly ash (50/50), *C:GBR* cement/ground burnt rock (50/50)

containing crushing screenings reinforced with cement is a control. It follows from the figure that, in the initial period of hardening, in compositions containing technogenic raw materials, and in comparison with the control, a slow set of strength occurs. In the future, the most intensive increase in strength is observed in compositions containing fly ash and metallurgical slag. The frost resistance grades of the experimental compositions were F25 and F30.

The data obtained are in accordance with the ideas about the composition of ash, slag, and burnt rocks as components that can exhibit latent hydraulic activity under certain conditions. This is one of the features of the fly ash and burnt ash studied. These additives do not have the ability to harden independently, being mixed with water, and they do not transform into a stonelike state. However, a mixture of such additives with some binders that contain lime in a free form or excrete it when interacting with water receives the ability to hydraulically harden. The activity of the waste is due to the presence of components with energy centers, the appearance of which is associated with the specifics of the formation of fly ash and burnt rocks of mine dumps.

During cooling, metallurgical slag crystallizes almost completely, and the glassy phase is practically absent in them. Steel slag dump is characterized by weak latent hydraulic activity. A feature of the studied compositions is the delayed formation of the structure in the initial period of hardening and a noticeable increase in strength, which continues for a long time (2 or 3 years). Compositions with ash and rock noticeably harden and in 90 days the strength increases by 2–3 times with respect to the strength of 28 days. The hardening process continues and can be during 2–3 years. However, the strength growth rate is reduced by this time.

Compositions with additives of ash (curve 2) by 90 days have the greatest strength compared with additives of ground burnt rock (curve 3) and slag (curve 4). Thus, the greatest ability to show pozzolanic activity is expressed in fly ash and to a lesser extent in slag. This conclusion is confirmed by the data of indirect estimation of the adsorption activity of the studied technogenic raw materials upon processing of its weighed sample with a dye solution (methylene blue—MB) and determination of the optical density of the filtered dye solution [76]. The adsorption activity in terms of the optical density of MB decreases in the series "ash-ground burnt rock-slag." The values of optical density, respectively, are 0.230, 0.183, and 0.164 [23, 29, 30].

Crushing screenings reinforced with inactive binders (curves 5 and 6) slowly gain strength and at the age of 90 days have a strength significantly lower than in compositions with ash, rock, and slag, but sufficient for use in the construction of road bases. It follows from this that it is more efficient to use ashes, burnt rocks, and steelmaking slags in the form of finely dispersed additives. Joint grinding them with cement is more expensive, and the strength is lower. Samples aged for 28 days had water absorption in the range of 1.5–3.1% and a coefficient of frost resistance of 0.83–0.90. The use of technogenic raw materials in the composition of mixtures for road surfaces allows us to save up to 30% of cement. The technology of arranging the base layers using technogenic raw materials practically does not differ from similar work with natural raw materials. The thickness of the compacted layer should be no more than 20 cm, and the largest grain size no more than 10 cm. This

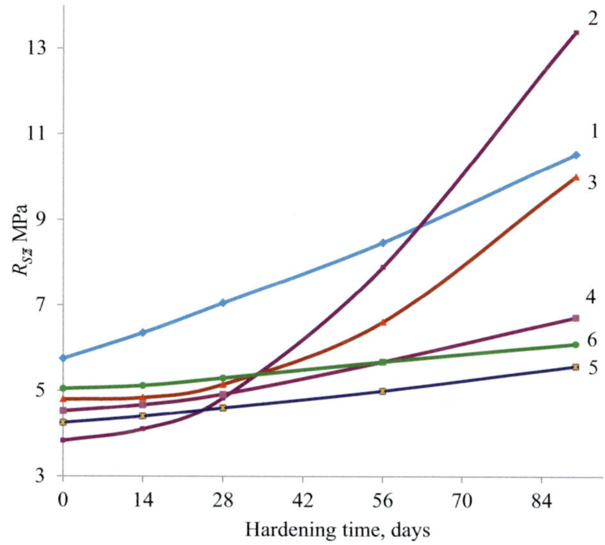

Fig. 5.5 Hardening kinetics of compositions containing technogenic raw materials: 1—control; 2—with fly ash; 3—with burnt rock; 4—with metallurgical slag; 5—with activated burnt rock; 6—with activated ash

ensures the achievement of the required strength and frost resistance and compliance of the physico-mechanical properties of the recommended compositions and mixtures with the requirements of regulatory documents for roads. It is very important that the bases are evenly and sufficiently compacted during construction and during operation of the road, and they do not allow bumps and subsidence that could lead to cracking and destruction of coatings.

In the underlying layers of the pavement bases, slag mixture, crushed stone, and burnt crushing screenings can be used to lay layers on roads of all categories, city streets, and roads (sidewalks and other landscaping elements) with an improved or transitional type of coating. Screenings of crushing of burnt rocks by grain composition correspond to a sand-gravel mixture. The sand fraction of screenings for crushing rocks in terms of physical and mechanical properties corresponds to sands of I or II class. They can be used as building sands. Based on modulo coarseness, sands are large and medium. Screenings are not subject to frost heaving, practically do not soak, and do not swell in water. By filtering ability, screenings are water permeable [119]. Filtering coefficient of screening of crushing changes from 3 to 11 m/day. According to their physical and mechanical properties, these screenings are suitable for use in road construction for the construction of subgrade, lining, drainage, and other layers of pavement. Direct screening can be used for the construction of the embankment and waterproof layers, for the strengthening of the subgrade, and as a granulometric additive to clay soils. Direct screening, as well as crushing screening, can be used in anti-silica layers. It is promising to use the potential of fillers from burnt mine rocks primarily for the construction of road pavement of III–IV categories of roads in the construction, reconstruction, and repair of roads of regional and local significance, and a large network of rural roads. The volume of such work is large. When using such raw materials, in addition to saving natural mineral raw materials, the cost of transportation for delivering raw materials to the place of consumption is reduced, and some environmental problems are solved to reduce the negative impact of the waste dump on the environment near its location.

5.3 Structural Layers of Road Pavements from Burnt Rocks of Mine Dumps

The development of effective road-building materials based on burnt rocks and ash-slag waste, capable of working under conditions of increasing loads and climatic influences, and technologies for the manufacture of structural layers of pavement of increased bearing capacity is relevant. The experience of using fly ash and ash-slag mixture in road construction is widely known. Below are the results of the use of burnt rocks of mine dumps in pavements from the example of the rocks of the Vorovskogo mine dump (Rostov region, Russia). This rock dump is formed by overburden formed during mining of coal seams located in the southern wing of the central part of the Shakhtinsko-Nesvetaevsky syncline. The roof of the worked-out coal seams is represented mainly by clay rocks, clay, sand-clay, slightly sandy, and carbonaceous schists, and less often sandstones. Subsequent transformations of stored rocks (weathering, thermal change, prolonged self-firing) led to significant changes in their mineral mass and physical condition. The fractional composition of the rocks composing the studied dump varies in a wide range, namely from blocks of 0.5–1.0 m in cross section to small crushed stone, gravel, and sand (see Fig. 5.6).

The crushed stone of various fractions from the rocks of this dump has a minimum crushing grade of 600 and a maximum grade of 1200, and a frost resistance of at least F50. Crushed stone withstands the tests for structural stability against all types of decay. The surface of the particles is not round, but rough. Aggregates do not contain clay and silt particles, and clay in lumps. Water resistance grade is W1 (weight loss after saturation with water and drying less than 1%). Figures 5.7 and 5.8 show the options for pavement, in the structural layers of which fractionated fillers from burnt mine rocks are used.

The experimental section of the technological road is being operated in the village Kiselevo, Rostov region (Russia). This is a section of the access road to the industrial site of the enterprise, made using materials from the burnt rocks of the dump of the former Kirov and Vorovsky mines, Rostov region, in the underlying and lower layers of the foundation. Industrial testing of the construction technology of foundations and coatings using materials from burnt rocks was carried out on experimental sections of technological and access roads, industrial sites, and sidewalks in the city of Gukovo (Rostov region). For this, burnt rocks of the dump of the former mine No. 26 were used. Table 5.10 shows the main properties of crushed stone fractions obtained with a two-stage scheme for the crushing of burnt rocks of the mine dump No. 26.

Tests of crushed stone from burnt rocks at the base of roads of categories IV and V with the installation of lightweight, transitional, and lower coatings were carried

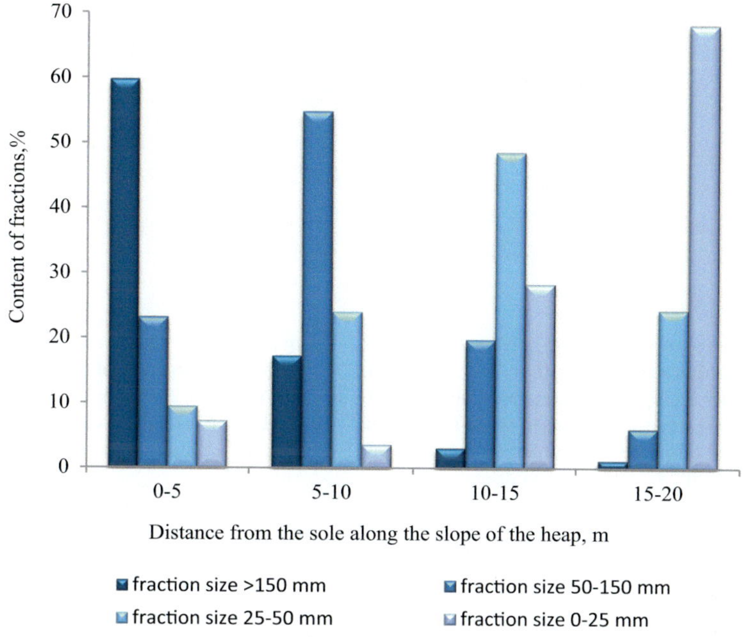

Fig. 5.6 Granulometric composition of the surface layer of the mine shaft

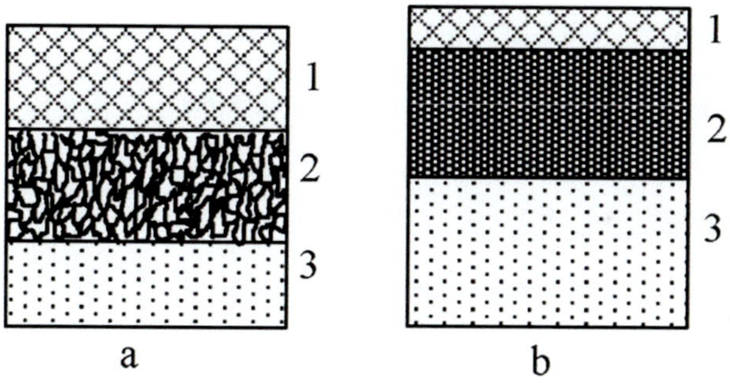

Fig. 5.7 Pavement design of the access roads of enterprises: (**a**) 1—fine-grained asphalt concrete, crushed stone fraction up to 15 mm, $h = 26$ cm; 2—crushed stone fraction of 5–10 mm, $h = 26$ cm; 3—sand, $h = 15$ cm; (**b**) 1—fine-grained asphalt concrete, crushed stone fraction up to 15 mm, $h = 5$ cm; 2—black crushed stone fraction of 5–10 mm, $h = 20$ cm; 3—sand, $h = 35$ cm

out at the mines "Almaznaya," "Zamchalovskaya," and "Vostochnaya." For the construction of the underlying layer, crushed stone mixture was used after the first crushing with sizes of grains from 20 to 150 mm. For the lower and middle layers of the bases, crushed stone from >40 to 80 mm and from >80 to 150 mm was applied, respectively; for the upper layers of bases and coatings, crushed stone from >20 to 40 and from >40 to 80 mm was applied, respectively. Crushed stone from burnt rocks is well compacted. Therefore, it is most efficient to stack it according to

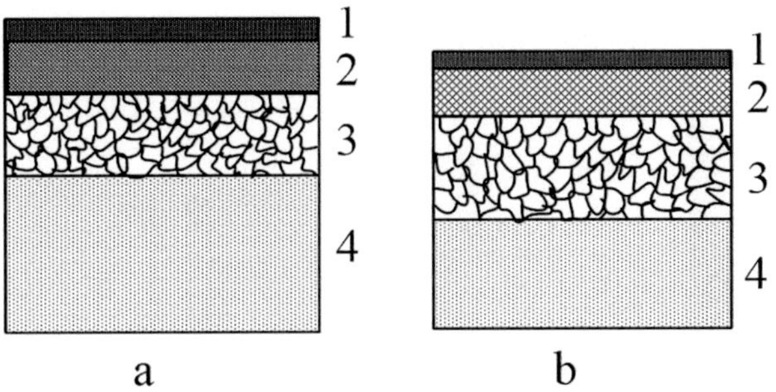

Fig. 5.8 Design of the road pavement improved lightweight type: (**a**) 1—surface treatment, crushed stone fraction of 5–10 or 10–15 (20) mm; 2–crushed stone fraction of 5–40 mm, treated bitumen, $h = 8$ cm; 3–crushed stone fraction of 40–80 (150) mm, $h = 20$ cm; 4—sand, $h = 26$ cm; (**b**) 1–surface treatment, crushed stone fraction of 5–10 or 10–15 (20) mm; 2–crushed stone fraction of 5–40 mm, treated bitumen on the road, $h = 8$ cm; 3–crushed stone fraction of 40–80 (150) mm, $h = 16$ cm; 4–sand, $h = 16$ cm

Table 5.10 Physical and mechanical properties of crushed stone from burnt rocks of mine dump No. 26

Indicators	Fraction, mm				
	From >5 to 10 mm	From >10 to 20 mm	From >20 to 40 mm	From >40 to 80 mm	From >80 to 120 mm
Volumetric weight, kg/m^3	1210–1290	1230–1260	1140–1170	1160–1190	1180–1230
Content of plate and needle-shaped grains, %	16.5–25.0	13.0–21.0	9.5–11.3	3.2–13.5	4.5–15.0
Grain content of weak rocks, %	2.5–8.0	3.1–7.7	1.5–6.8	1.3–5.2	1.1–4.6
Strength, grade	1000–1200	1000–1200	800–1200	800–1200	800–1200
Frost resistance, grade	F50	F50	F50	F50	F50
Abrasion, grade	A1	A1	A1	A1	A1
Resistant to all types of decay, weight loss, %	0.4–0.9	0.7–1.1	0.3–0.6	0.3–0.7	0.5–1.3
Water resistance, weight loss, %	W1; W2	W1; W2	W1; W2	W1; W2	W1; W2

the method of optimal mixtures. In the subsequent layers of bases and coatings, crushed stone obtained by re-crushing burnt rocks should be used. Optimal crushed stone mixtures consist of successively decreasing fractions: 5–10, 10–20, and 20–40 mm. It could also be used as a mixture of fractions: from 5(3) to 20, from >0 to 20, and from >0 to 10 mm.

The technological process of the building layer was reduced to leveling the rock and compaction of the planned surface with light or medium rollers with metal rollers. The resulting subsidence was leveled, scattering burnt rock of smaller size. Subsequently, with a maximum layer thickness of up to 20 cm (in a loose state), heavy rollers were passed. The total number of passes of the roller is 15–18. To obtain a dense and durable monolithic crushed stone layer, it is necessary to ensure the correct watering regime. At the beginning of compaction in loose placer, individual grains of crushed stone are easily distributed and mutually displaced. At the same time, large fractions of crushed stone play the role of a kind of spatial framework. Small particles fill the voids of the frame. During this period, compaction can be done without watering. When the initial sedimentation of the layer is achieved, further compaction requires overcoming friction between the particles. Water in this case facilitates compaction, and also moisturizes fines and dust from well-baked rocks, resulting from breaking of the edges of rubble during rolling. Final compaction to a given density is carried out without watering. Over time, moistened dust cements and strengthens the structure of the crushed stone layer. Estimated water consumption is 12–20 L/m^2. In the dry season, the water flow for irrigation is 4–5 L/m^2. With a lack of water, the compaction time is lengthened; with an excess, the underlying layer can be overwetted and gravel rolling is complicated due to the adhesion of cement paste and individual gravels to the rollers of the roller. Compaction was completed when the deformation stopped after the passage of a heavy roller, and the rock density reached 1.85–1.90 g/cm^3. In the subsequent layers of bases and coatings, crushed stone obtained by re-crushing burnt rocks should be used. The compaction coefficient is 1.30–1.35.

If the gravel mixture of the optimal grain composition is used for the construction of pavement layers, then the material is immediately distributed with a layer of the required thickness, leveled with a grader or bulldozer or compacted. The number of roller passes is defined experimentally. The compacted layer can reach a thickness of 20 cm (in a loose body), but this requires at least 15–18 passes of the heavy roller in one place. With significant porosity of areas of the compacted layer, fines of 5–10 mm in size are scattered on them and additionally compacted with 1–2 passes of a heavy roller.

Testing of crushed stone from burnt rocks in road construction was carried out by comparing the condition of the experimental sections of the road at certain intervals of time of their operation (twice in 3 years). From the experimental sections of the road, crushed stone samples were selected to test at different times of operation of the road. Test sections were prepared in various ways.

In one of the sections, the crushed stone mixture from >20 to 120 (150) mm was distributed on a dirt road. The length of the section was 650 m, and the stacked volume of gravel was 847 m^3. The thickness of the layer was 15 cm. The access road was designed for movement in the main cargo transport, and for transportation of mine rock. Crushed stone distribution was carried out without preparing the subgrade, and without planning and compaction. The compaction of crushed stone occurred spontaneously under the influence of a moving vehicle. The lack of lateral emphasis did not allow one to maintain the contours of pavement. The thickness of the crushed stone layer along the road section was uneven. The surface layer was unstable. Periodically, it is necessary to restore it by returning the grader to the carriageway of the material pushed away by the wheels and adding a new one.

On a section of the road with a length of 120 m (see Fig. 5.9) at the Almaznaya mine, the volume of the laid rock is 205 m^3, and the thickness of the base is 20 cm,

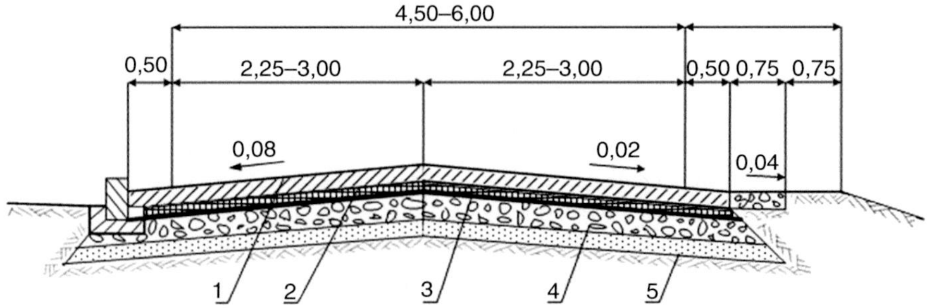

Fig. 5.9 Constructive transverse profile of the road with a width of the carriageway of 4.5–6.0 m; pavement design: 1—medium-grained asphalt concrete, $h = 4$ cm; 2—coarse-grained asphalt concrete, $h = 5$ cm; 3—burnt rock crushed stone impregnated with bitumen, $h = 6.5$ cm; 4—base of burnt rock rubble, $h = 20$ cm; 5—sandy underlying layer, $h = 10$ cm

which is made of fractioned gravel: the bottom layer from the fraction from >20 to 80 mm, and the upper layer from the fraction from >5 to 20 mm with bitumen impregnation. A two-layer coating of asphalt concrete is applied to the base: the upper layer is fine-grain and the lower layer is coarse-grain asphalt concrete. The strength and particle size distribution of crushed stone from the base of the road with asphalt concrete pavement did not change during the operation of the road. Fraction of crushed stone from >20 to 80 mm has a grade of strength of 800, and the fraction from >5 to 20 has a grade of strength of 1000.

At one of the experimental sections of the access road, a decrease in the crushed stone strength for the fraction from 10 to 20 mm on one grade was stated. The strength of this fraction was 800. The decrease in strength was associated with a partial change in the particle size distribution of this fraction.

In another experimental section of 400 m long, the gravel mixture from 20 to 120 (150) mm was laid on the on-site dirt road. The volume of crushed stone laid was 2404 m^3, and the layer thickness was 40 cm. The crushed stone mixture was leveled and compacted with rollers. The surface of the densified layer is relatively flat and uniform in density. There are no deep traces from the wheels of transport, loosening, and subsidence of gravel.

In sections of the lower type of road (Vostochnaya and Zamchalovskaya mines), no significant changes in crushed stone strength occurred. The results of the examination of the condition of the experimental sections of the road show that on well-compacted sections of the road the deviation in the width of the structural layer, the road slopes, and the height of the base do not exceed the permissible limits; there were no significant changes in the strength of crushed stone. The condition of the experimental sections of the road is quite satisfactory. Loosening and planting of crushed stone were not observed. One of the advantages of using crushed stone from burnt rocks in road construction is that due to its light compatibility and cementing ability of the dust fraction, dense packing of aggregate grains is achieved and a monolithic structure of increased strength and stability is created. The bases, arranged by the method of optimal mixtures, have a strength sufficient for construc-

Table 5.11 Compositions based on ash-slag waste

Composition number	Mixture composition, weight %					
	Slag mixture	Fly ash	Cement, M400	Sodium hydroxide	Liquid glass	Calcium chloride
1	96.0	–	4.0	–	–	–
2	94.0	–	6.0	–	–	–
3	97.0	1.5	1.5	–	–	–
4	96.0	2.0	2.0	–	–	–
5	60.0	38.8	–	1.2	–	–
6	86.0	–	–	–	11.0	3.0
7	89.0	–	–	–	9.0	2.0
8	77.0	–	1.1	–	17.0	5.0
9	77.0	–	0.5	–	18.5	4.0

Table 5.12 Physico-mechanical properties of compositions for road surfaces

Composition number	Compressive strength (MPa) of water-saturated samples at the age of days				Bending strength (MPa) of water-saturated samples at the age of days		Compressive strength (MPa) of water-saturated samples after 50 cycles of freezing/thawing at the age of days	
	7	28	90	180	28	90	28	90
1	2.8	3.5	5.3	6.9	0.7	1.1	2.55	3.87
2	3.7	5.1	8.7	10.8	0.8	1.3	3.77	6.44
3	2.7	4.6	9.4	11.7	1.3	1.8	3.31	6.77
4	3.0	4.9	10.0	12.3	1.5	2.9	3.67	7.50
5	7.4	8.0	13.2	15.8	3.7	4.2	6.16	10.16
6	5.2	7.7	12.8	13.0	2.3	3.5	5.77	9.60
7	4.7	7.3	10.9	11.1	1.9	3.3	5.40	8.07
8	8.0	12.3	17.0	17.3	2.9	4.0	9.25	12.58
9	6.8	11.0	14.7	14.9	2.2	3.7	8.14	10.88

Note: Liquid glass with a module of 1.5–2.0 and a density of 1.25–1.30 g/cm^3; mixing of mixtures is carried out to the optimum moisture content of maximum compaction

tion of the overlying layers of pavement. Positive practical experience in the use of crushed stone and sand mixtures from burnt rocks in road construction confirms the suitability of these materials for laying underlay layers and road bases. Qualitative fillers from burnt rocks can be used in all layers of pavement.

For the construction of roads of local importance, compositions of ash-slag waste from Novocherkasskaya TPP with activating additives were developed. The compositions are given in Table 5.11 and the characteristics of their properties are in Table 5.12.

The results show that the optimal composition of the mixtures and their properties depend on the amount and composition of the binder. In pavements, they can be used as load-bearing and at the same time frost-protective lower and upper layers of the bases of roads of all categories, with the exception of the upper layers of roads of category I. On the roads of categories IV and V, these compounds can be used for coatings with the obligatory installation of a wear layer on them.

From the data of Table 5.12, it follows that the additional activation of slag by binders leads to an increase in the physical and mechanical parameters of mixtures. The greatest reactivity is provided by grinding (0.5–1.0 h) with caustic soda additives (composition No. 5), during which chemical and mechanical activation occurs. High values of physico-mechanical properties are observed in compositions reinforced with liquid glass. However, the softening and frost resistance coefficients for these compositions are lower compared to composition No. 5.

The softening coefficient for hardened slag mixtures is in the range of 0.83–0.95, and frost resistance is 0.75–0.74. The greatest results were noted for composition No. 5. To impart water resistance to mixtures with liquid glass in the initial period of hardening and under adverse weather conditions, they are treated with a solution of calcium chloride. The value of the calculated modulus of elasticity of laboratory samples of hardened slag mixtures is in the range of 300–650 kg/cm^2.

Unreinforced ash-slag mixtures can be used for the construction of road bases independently and mixed with coarse gravel. The construction of structural layers of pavement using unreinforced slag mixtures does not differ from the building of similar layers from natural finely detrital materials. Ash-slag mixtures with binders are prepared using a stationary mixing unit. In this case, the operations are performed in the following sequence: the slag mixture is developed in dumps, loaded with an excavator into dump trucks, and delivered to the place of preparation of the mixture. The dosed components of the mixture (slag mixture, additives, and water) are fed into the mixer and mixed for 1–1.5 min until the components are distributed evenly and uniformly. The prepared mixture is transported to the road for laying. The mixture can be prepared directly on the road using mobile mixing plants and/or road milling machines. For a uniform distribution in the mixture, additives of binders are more effective to introduce mixing with water. The mixture is moistened to the optimum sealing moisture and mixed with a motor grader with a disc harrow or a road mill, and then the mixture is distributed with a motor grader over the entire width of the base and compacted with rollers.

Reinforced ash-slag mixtures in the initial period of hardening gain strength slowly, and therefore a time gap from the moment the mixture is prepared to laying is allowed up to 7–8 h. If necessary, to accelerate the hardening processes, the densified bases from slag mixtures are impregnated with a calcium chloride solution of density 1.15–1.18 g/cm^3.

During the experimental construction of a 1.5 km long section of a local road, the results of laboratory tests were checked and techniques for production were worked out. For the construction of the road, slag mixture was used. An asphalt coating of 3 cm thick was applied to the base. The operation of the experimental section of the road for many years confirms the practical feasibility and effectiveness of the use of slag mixtures for construction of the lower base layers in local roads.

5.4 Asphalt Concrete Coatings Using Technogenic Raw Materials

This section presents the results of using the burnt rocks of the Mayskaya and No. 26 mine dumps located in the Rostov region (Russia), and the influence of their specific features on the formation of the structure and properties of organic mineral composites (asphalt concrete). Table 5.13 shows the chemical composition of the rocks of these dumps.

The burnt mineral part of asphalt concrete mixtures and asphalt concrete was prepared at crushing and screening complexes at the Mayskaya and No. 26 mine dumps. To obtain crushed stone and screenings from burnt mine dumps, burnt and re-burnt mine rocks, represented by mudstones, siltstones, and sandstones with compressive strength in a water-saturated state of at least 600 kg/cm^2, are suitable. To prepare the mineral component of asphalt mixtures, products after the second crushing stage were used. Depending on the fractional composition, fillers from burnt mine rocks are characterized by the following indicators of physical and mechanical properties:

(i) Grade of crushed stone by strength during the crushing test: 600, 800, 1000, 1200
(ii) Grade on abrasion: A1, A2
(iii) Frost resistance grade: F25, F50, F100

Crushed stone is resistant to all types of decay, and does not contain clay, silty particles, and clay in lumps; it withstands water resistance tests.

The gravel contains particles of a cuboid shape, as well as lamellar and needle shape. However, the content of the latter does not exceed regulatory requirements. The presence of needlelike particles creates a reinforcing effect when laying crushed stone mixtures. The surface of the particles is not rolled, torn, rough, and clean; that is, clay and other contaminants are not contained on the surface of the particles. This feature of such fillers increases the adhesion of gravel particles in the structural layers of pavements and with binders of organic origin. As a result of thermal exposure, the rock particles acquired a porous structure. However, the low values of water absorption indicate that most of the pores of the gravel particles are closed. The presence of porosity is associated with a lower bulk density in comparison with natural gravel, increased sorption ability with respect to bitumen, and improved thermal insulation properties of base layers with such fillers. The content of carbonaceous impurities does not exceed 3%. Screenings of crushing of burnt rocks by

Table 5.13 The chemical composition of the burnt rocks of mine dumps

Mine	SiO_2	Al_2O_3	Fe_2O_3	CaO	MgO	TiO_2	K_2O+ Na_2O	SO_3 Com.	IL	M_{cg}
Mayskaya	59.73	19.22	5.87	1.29	2.69	1.00	3.00	2.78	4.40	0.42
No. 26	56.65	21.79	6.60	1.63	2.07	0.96	2.98	2.71	4.94	0.50

Note: *IL* ignition loss, M_{cg} is the clay-glandular module, equal to $(Al_2O_3 + Fe_2O_3)/SiO_2$

Table 5.14 Grain composition of the mineral part of mixtures for upper layers of coatings, %

Kind and type of mixtures and asphalt concrete	Grain sizes, mm, no more									
	20	15	10	5	2.5	1.25	0.63	0.315	0.16	0.071
Dense B	100	85	70	55	38	29	20	14	9	6
	100	88	81	60	46	33	24	17	13	7
	100	96	86	65	52	38	27	22	16	10
Dense G	–	–	100	95	68	45	28	18	11	8
	–	–	100	97	77	56	39	26	19	12
	–	–	100	98	85	67	50	35	24	16

grain composition correspond to a sand-gravel mixture. The sand fraction of screenings for crushing rocks by physical and mechanical properties can correspond to sands of I or II class.

In asphalt concrete mixtures intended for the upper layers of road surface coatings of IV and V road categories, continuous-grain burnt rock fillers, prepared from fractions from >5 to 15 mm and screenings of crushing from 0 to 10 mm, were used. Screenings of crushing by grain composition (from 0 to 10 mm) were a mixture of sand and gravel fractions. The content of the latter ranged from 25 to 30%. A fine screening fraction (0–0.071 mm) served as a mineral powder in the composition of asphalt concrete. Preparation of mixtures for the study was carried out in laboratory conditions, and for experimental sections of roads and sidewalks. Experimental mixtures of an asphalt concrete plant were used. For the tests, the compositions of the fine-grained dense asphalt concrete of type B with grade II and the dense sandy of type G with grade II were selected. Table 5.14 shows the grain composition of the mineral part of the studied asphalt mixtures for dense asphalt concrete of types B and G. The bitumen content was from 7.0 to 9.0%.

When selecting the composition of asphalt concrete, the optimum amount of bitumen was calculated based on the residual porosity established for this type of asphalt concrete, and was refined during trial batches according to the results of strength. Under laboratory conditions, an experimental determination was made of the adhesion of the binder film to gravel particles after they were treated with bitumen according to the method of SoyuzdorNII [36]. Adhesion was evaluated visually by the surface area of crushed stone previously treated with bitumen, on which a bitumen film was preserved after the samples were kept in boiling distilled water. The results showed a good adhesion of bitumen to rock-crushed stone: at least 95% of the area of the gravel is covered with a film of binder.

Mixture preparation and testing of samples were carried out in accordance with the requirements of regulatory documents on testing methods for asphalt concrete mixtures and asphalt concrete. The properties of asphalt concrete were tested on cylindrical samples with a diameter and height of 50.5 mm, molded on a press with a bilateral application of a compressive load of 40 MPa. From experimental asphalt concrete mixtures taken at the asphalt concrete plant and in the experimental sites, cylindrical samples with a height and a diameter of 71.4 mm were prepared.

The technology for the preparation of asphalt concrete mixtures with burnt rocks did not fundamentally differ from that adopted at the plant. Only an increase in the viscosity and stiffness of the mixture was noted. This is due to the shape and roughness of the crushed particles. A significant part of the grains of the mineral material has an acute-angled shape, and a rough surface, which makes the mixture stiff. Increased viscosity makes mixing difficult. Additional efforts are required for mixing and uniform distribution in the entire volume of the components of the asphalt concrete mixture during its preparation. Moreover, it is necessary to increase the consumption of a binder for fine-grained asphalt concrete by 2–3% compared with the consumption of bitumen for asphalt mixtures on natural stone materials. However, this is not an obstacle to the use of materials from burnt rocks in asphalt mixtures. It is only necessary to adjust the technological methods for the preparation of asphalt mixes, for example, to select plasticizing additives and the optimum temperature of the mixture, adjust the mixing mode of the components, and improve the particle size distribution of the fillers. To evenly distribute and envelop the surface of the mineral part of the mixture, the mixing time was increased by 10–15%.

It is possible to reduce the rigidity of the asphalt concrete mixture and reduce bitumen consumption by introducing finely ground additives, for example, ash from thermal power plants, and burnt rock. A good effect is achieved when up to 30% sand is added to the mixture. Possessing hydraulic activity and high dispersion, these additives in asphalt concrete will fulfill not only the role of plasticizer, but also the function of a fine microfiller. Ashes and ground burnt rocks can replace traditionally used mineral powder. Asphalt concrete containing active mineral fillers from ash or ground mine rock is highly resistant to frost. Fly ash and ground mine rock can be recommended as a mineral powder in asphalt concrete for paving, not only in regions with a temperate climate, but especially in harsh climatic conditions and in areas of high humidity. This conclusion is applied to ash-slag waste. The test results for experimental asphalt concrete are given in Table. 5.15.

Based on the experimental compositions of asphalt concrete mixtures, repair works were carried out at the industrial site and the pavements and road sections of the industrial facility in the Gukovo city of Rostov region (Russia) were covered. The condition of the experimental sections of roads, sidewalks, and industrial sites was monitored for 3 years from the date of their manufacture. During field observations, the evenness of the coating and the state of the surface (the presence of shells, sagging, cracks, quality of the joints, etc.) was monitored. Every year, core samples were drilled from the test sections in the form of cuttings 50 × 50 cm^2 in size to determine the time-dependent changes in the physical and mechanical properties of asphalt concrete during operation. The results of such determinations are given in Table. 5.16. When sampling, the thickness of the layers was measured and their adhesion to each other and to the base was visually evaluated. Comparison of the physico-mechanical properties of cores and reformed samples indicates that the performance characteristics of asphalt concrete during compaction in operation are improved.

Table 5.15 Physico-mechanical properties of asphalt concrete based on burnt rocks

Indicators	Values of indicators for asphalt concrete	
	Laboratory	Factory
Average density of asphalt concrete, g/cm^3	2.39–2.41	2.23
Average density of mineral part, g/cm^3	2.00–2.10	2.07
True density of mineral part, g/cm^3	2.63–2.68	2.73
True density of asphalt concrete, g/cm^3	2.37–2.48	2.35
Porosity of mineral part, %	21.8–23.3	17.0
Residual porosity of asphalt concrete, %	2.7–3.1	2.9
Asphalt concrete water saturation, %	1.38–2.16	3.3
Asphalt concrete swelling, %	0.44–0.51	0.48
Compressive strength of dry samples, MPa:		
At temperature 50 °C	1.8–2.7	1.4
At temperature 20 °C	5.5–8.4	6.9
At temperature 0 °C	12.4–13.0	10.1
Compressive strength of water-saturated samples, MPa: At temperature 20 °C	5.45–8.8	6.2
Compressive strength of the samples with prolonged water saturation, MPa:		
At temperature 20 °C	3.85–9.4	6.0
Water resistance coefficient	0.99–1.05	0.9
Water resistance coefficient with prolonged water saturation	0.7–1.19	0.87
Adhesion of bitumen to the surface of the mineral part of asphalt concrete	Excellent	Excellent

For asphalt concrete based on the materials from burnt rocks, high values for strength, water resistance, frost resistance (see Figs. 5.10 and 5.11), and low values for water saturation were revealed. This is due to the good adhesion of the active grains of filler to an organic binder.

In domestic practice, there is experience in the use of mineral industrial waste and, in particular, burnt rocks for the manufacture of asphalt concrete mixtures. So, scientists of Tomsk University [98] studied the fatigue properties of asphalt concrete from mineral industrial waste: burnt rocks of mine heaps as a large mineral filler, TPP slags as a sand filler, and activated ground slags as a mineral powder. The dense asphalt concrete has best fatigue properties. With an increase in porosity, the stress concentration at the pore surface increases, which leads to more rapid formation of microcracks and destruction of the material. This indicates the need for high-quality compaction of asphalt concrete pavements during their construction in order to increase the service life of pavement. The asphalt concrete endurance is largely influenced by the degree of the adhesion of binder to the surface of mineral fillers. The test results obtained by the authors indicate good adhesion of bitumen to burnt rocks. It is shown that fine-grained asphalt concrete with aggregates from burnt rocks in terms of endurance to cyclically repeating mechanical loads does not only concede to the standard samples manufactured from crushed stone, sand, and mineral powder, but also surpass them. The positive test results of mineral materials

Table 5.16 Properties of corns-cuttings and samples of asphalt concrete pavements

Corns-cuttings					Reformed samples				
Average density, g/cm^3	Water saturation, % by volume	Compressive strength, MPa (at $T = 20$ °C)	Water resistance	Water resistance at prolonged water saturation	Average density, g/cm^3	Water saturation, % by volume	Compressive strength, MPa (at $T = 20$ °C)	Water resistance	Water resistance at prolonged water saturation
2.39	2.15	8.8	1.11	0.96	2.23	1.07	8.2	0.98	0.99
2.41	1.93	9.1	1.18	1.03	2.32	1.05	8.5	0.96	1.10
2.42	1.87	9.7	1.27	1.09	2.30	1.07	8.8	1.01	1.13

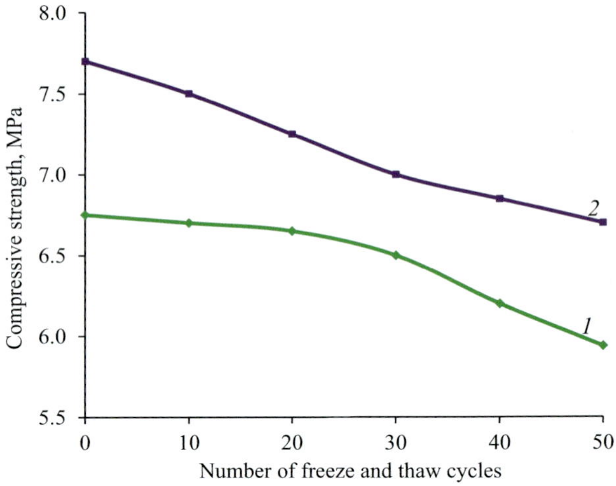

Fig. 5.11 Change in the compressive strength of asphalt concrete on the duration of test: 1—type B; 2—type G

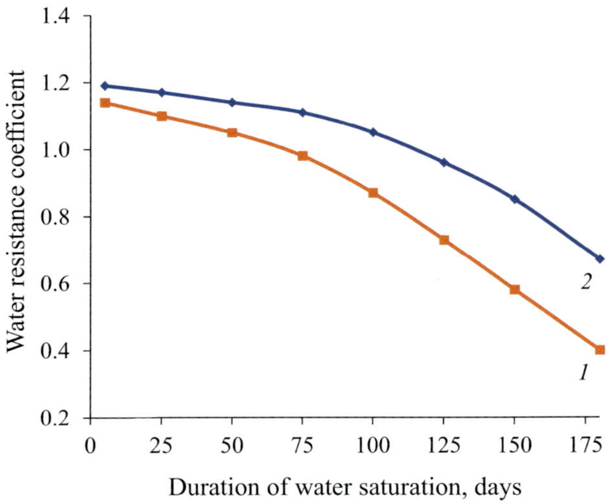

Fig. 5.10 Change in the coefficient of water resistance of asphalt concrete with prolonged water saturation: 1—type B; 2—type G

from burnt rocks of mine dumps confirm the possibility of their use in asphalt concrete mixes and asphalt concretes for the construction of the lower layers and coatings of roads of IV and V categories. By using burnt mine rocks in road construction, it is possible to improve the quality of asphalt concrete, extend its service life by improving physical and mechanical properties, reduce the consumption of natural resources, and use cheaper road-building material.

5.5 Concrete Coatings Using Technogenic Raw Materials

One of the measures to improve the ecological state of the human environment is the replacement of asphalt concrete pavements of sidewalks, sites, and pedestrian walkways with small-sized concrete products based on mineral fillers and binders. Fine concrete products must have high strength, frost resistance, wear resistance, and durability. Obtaining products with such characteristics from ordinary cement-sand concrete is not always possible. Natural sands often have a low modulus of fineness (for example, in the Rostov region, it is equal to 1.2–1.4). According to the requirements of regulatory documents, the use of such sand in concrete is not allowed, since this inevitably leads to a significant cost overrun of cement. Therefore, for fine-grained concrete, other fillers should be recommended, ensuring the achievement of the necessary characteristics of concrete within the established norms of cement consumption.

Studies and practical experience have proved [20, 36, 38, 166] that high-strength fine-grained concrete can be obtained on the base of slag waste and screenings of burnt mine rocks. As it is well known, the slag mixtures from hydraulic dumps contain a fine-dispersed fraction in the form of fly ash and fuel slag. Fly ash is a dispersed powder consisting of tiny spherulites ranging in size from 0.001 to 0.14 mm. Fuel slag consists mainly of grains from 0.14 to 10 mm in size. Therefore, the slag fraction of 0.14–2.5 mm used in concrete can play the role of a fine filler, and a fraction of 3–10 mm to be a large filler. Thus, the slag mixture located in the dumps is a composition consisting of three components: slag crushed stone, slag sand, and pulverized fly ash. Each of these components in concrete performs its function; therefore, the proportion of each of them in the slag mixture and the ratio between them are of great importance. The properties of these components are such that they allow the use of slag raw materials for the manufacture of concrete for various purposes.

The granulometric composition of the slag mixture from the dumps is close to optimal. The mass fraction of grains of various sizes in its composition is in the following ranges: 3–10 mm (42–55%); 0.14–2.5 mm (22–40%); and fraction finer than 0.14 mm, that is, fly ash (18–25%). The best physical and mechanical properties of fine-grained concrete were obtained using slag mixture with a particle size modulus of 2.2–2.8. Thus, the use of slag with a particle size modulus of 2.65 in concrete compositions for paving slabs allowed saving from 100 to 250 kg/m^3 of cement.

Another effective filler in fine concrete for road construction is screenings for crushing burnt mine rocks. By processing burnt rocks into fillers, the yield of screenings is about one-third, and there is no shortage of these materials. One of the features of screenings for crushing burnt rocks is an increased content of grains of flaky and needle shape. The increase in the content of flaky grains in the screening increases water demand and voidness, and reduces the mobility of the concrete mixture. An important characteristic of screenings is their size. This characteristic has the greatest impact on the rheological properties of concrete. The modulus of coarse screening, checked for a long time, varies from 3.3 to 3.7 with an average value of 3.55. The screening does not contain clay and other clogging impurities and harmful inclusions. Large screening fractions have a strength grade of at least 800.

Studies have shown [18, 20, 38] that based on screenings of crushing burnt rocks, concrete of classes B15–B22.5 can be obtained. The consumption of crushing screenings was 1300–1500 kg/m^3, and cement 250–350 kg/m^3. To improve the technological properties of the concrete mixture and increase the density of concrete, sand was additionally introduced in an amount from 15 to 27%. The same effect on the properties of the concrete mixture and the quality of concrete is provided by the addition of direct (original) screening of burnt rocks. This conclusion is of great practical importance, since in many areas of the Rostov region there is a shortage of sand, but screenings of burnt rocks (both original and after crushing) are in abundance. To reduce the rigidity of the concrete mixture, fly ash or fine-ground burnt rock is introduced.

Most dense and strength concretes of classes B22.5–B40 were obtained by vibropressing technology from concrete mixtures with a reduced water-cement ratio (up to 0.25). Good workability of such stiff concrete mixtures was achieved with the introduction of dry ash. This additive allows one to reduce cement consumption by 100–150 kg/m^3, voidness by 7–13%, and water demand by 5–7%.

Studies have shown that the strength of fine-grained concrete based on slags and screenings of burnt rocks with finely ground additives is affected by the regime of heat and moisture treatment. In particular, the temperature of isothermal heating should be at least 80–85 °C. A decrease in the temperature of isothermal heating from 80 to 65 °C leads to a significant decrease in the strength of concrete; an increase in temperature to 95 °C is accompanied by a slight increase in strength. Therefore, the following mode can be considered optimal: holding the freshly formed products for at least 2 h, raising the temperature before isothermal heating for 3 h, holding at the temperature of isothermal heating during 8 h, and cooling the goods for at least 3 h.

Some authors recommend increasing the temperature of isothermal heating by 10 °C and the exposure time by 2 h in order to use more completely the reactivity of ash and burnt rock. However, in practice, in factory conditions, the temperature of isothermal heating, as a rule, is equal to 60–80 °C, and in winter conditions the temperature can be even lower. The unused reserve of reaction ability of ash and burnt rock is realized at the subsequent increasing strength for a long time. The strength properties of ash-concretes of 200 and 300 grades for compressive strengths were tested. These samples were stored for 2 and 6 years at a temperature of 18–20 °C and relative humidity of 55–65%. After 2 years, the strength increased 1.5–1.7 times, and after 6 years 2.0–2.6 times. Strengthening of concrete with hydraulic additives can also occur under normal hardening conditions with a relative humidity of 90–100% and a temperature of 15–20 °C. The formation of the structure of cement stone under these conditions occurs during heat release during the hydration of the cement. However, the greatest effect of the use of additives exhibiting pozzolanic activity is achieved with heat-moisture treatment.

Samples of burnt rock concrete on the base of screenings from the rocks of the dumps of mine No. 26 and Petrovsky mine, after hardening for 28 days under normal conditions, were left (i) in water, (ii) in an aggressive sulfate medium, and (iii) in air at a temperature of 12–15 °C and humidity of more than 70%. After a year and a half, the samples were tested and showed an increase in strength in water of 35–40%, in wet conditions of 30–35%, in air of 17–25%, and in an aggressive environment of 15–20%.

Concretes containing industrial raw materials (ash-slag waste or burnt mine rocks) have high bending strength. Figure 5.12 shows the dependence of bending strength on compressive strength for three types of concrete: ash concrete (curve 1), rock concrete (curve 2), and concrete on the base of conventional materials (control curve 3). The ash concrete has greatest values of bending strength, and the smallest values correspond to ordinary concrete.

The increased strength of ash concrete is associated with a pronounced plasticizing effect of ash, which contributes to the formation of a more uniform and dense concrete structure. Ash more actively than burnt rock participates in the physicochemical processes of concrete hardening, and as a result, the amount of cementitious substances in the ash concrete is greater than that in the rock and ordinary concrete. Another distinctive feature of concrete on the base of technogenic raw materials is increased sulfate resistance. Observations of the expansion of samples upon alternating saturation in a solution of sodium sulfate and drying show that in ash concretes and burnt rock concretes, the deformations, comparable in magnitude to the deformations of ordinary concrete, appear with a significantly larger number of test cycles. In Fig. 5.13, the curve for ordinary concrete is located to the left and has a steeper rise than curves for concrete with ash and burnt rock. The effect of ash

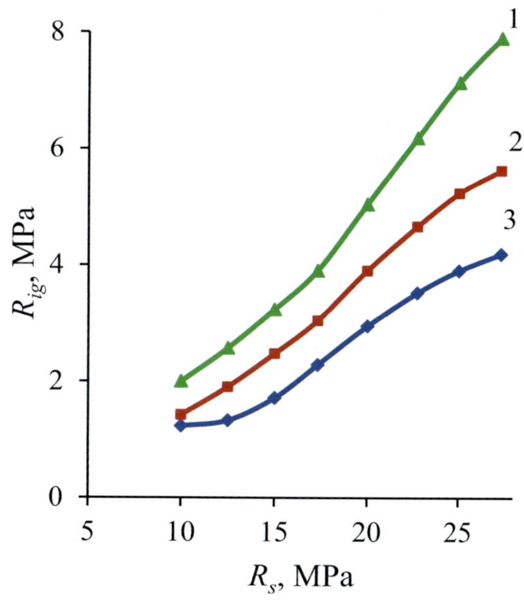

Fig. 5.12 Concrete compression strength dependence: 1—ash concrete; 2—burnt rock concrete; 3—ordinary concrete (control)

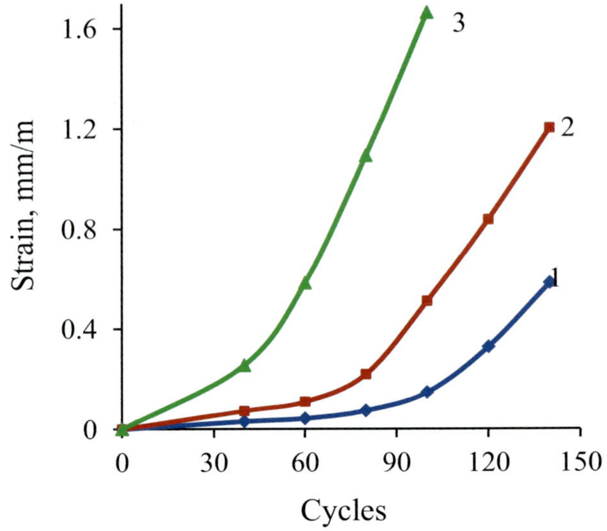

Fig. 5.13 Kinetics of the development of linear strains of concrete during cyclic tests in a solution of Na_2SO_4: 1—composition with the addition of burnt rocks; 2—composition with the addition of ash; 3—control composition

and burnt rocks on the sulfate resistance of concrete has a physical and chemical character. The physical nature of this drying is associated with a relative increase in contacts between particles and the formation of a more uniform and dense concrete structure. The physicochemical nature is due to pozzolanic properties, which are able to exhibit ash-slag waste and burnt rocks. The result of the manifestation of pozzolanic activity is the binding by mineral additives of free lime formed during the hydration of cement and the appearance of a significant amount of cementitious substances. In this case, the contact layer has a more developed and less defective surface, more resistant to deformations and aggressive media.

Climatic features of any country stipulate high demands on the physico-mechanical properties of road goods. For the production of high-quality paving slabs and curbs, a high degree of compaction of the concrete mixture is required. The effectiveness of the use of fine-grained concrete increases in the manufacture of small-piece products from them by the method of vibrocompression (vibro-stamping). This method of manufacturing small-piece products is provided in the technological lines of firms: Besser (USA), Hess, Henke (Germany), Longinotti, Rosacometta (Italy), and others. Using one line, using simple and easy labor-consuming equipment (molds) readjustment, it is possible to produce various types of goods: original wall blocks, self-blocking stone blocks for paving, facing slabs, road curbs, curly paving tiles, and other goods. Production of goods with a textured layer adding dyes and with the texture of natural stone is possible. The domestic lines of the Rifey series have proven themselves well. The line "Rifey-universal" has the following advantages: simple operation and maintenance; good maintainability due to convenient access and ease of disassembly of each unit; use of the same type and non-deficient components; high compactness; and ability to quickly change lines to another type of good by changing matrices. The line allows one to use almost any domestic materials.

The technological process of manufacturing products on such lines consists of the following basic operations: (i) dosing fillers and cement and feeding them into a forced-action mixer with automatic dosing of water; (ii) supply of concrete mixture through a conveyor tape from a mixer to a molding machine with a vibrator on; (iii) pressing the mixture in the mold with vibration for 6–7 s to the specified parameters; (iv) pushing goods out of the mold and lowering them onto pallets; (v) feeding pallets with freshly molded products to a multistory elevator; (vi) filling the elevator and feeding the goods into the chamber for heat treatment; (vii) transportation of finished goods for packaging; (viii) emptying, cleaning, and returning pallets; and (ix) transportation of finished stacks to a warehouse and loading. All technological operations are performed in automatic regime. Adopted technology allows one to get goods that meet existing standards and requirements, using the minimum amount of cement.

Based on the recommended compositions of fine-grained concrete under the conditions of production, an experimental batch of paving slabs using the technology of volume vibropressing was produced. From the products of paving slabs, cubes were made and tested. Physico-mechanical properties of fine-grained concrete paving slabs of the experimental batch are given in Table 5.17.

Paving slabs have a concrete class of compressive strength B30 and B40 (grade 400 and 500), frost resistance: F200 and F300 practically without reduction in strength. The mass loss during the wear test does not exceed 0.9 g/cm^2; water absorption of the tile is equal to 2.2–3.9%. Concrete grade on water resistance is W12. Figure 5.14 shows the microstructure of fine-grained concrete using technogenic raw materials. Materials from ash-slag waste and burnt rock participate in the hardening of cement with the formation of low-basic hydrosilicates, which have increased strength and durability. As a result, all the potential possibilities of cement are involved, and its utilization factor increases. By hardening such a composition, an optimal dense and more durable structure is created. This contributes to the production of dense, high-strength, and frost- and wear-resistant concrete with a reduced consumption of cement clinker.

The results of acceptance tests of the experimental batch confirm the feasibility of using ash-slag waste and burnt mine rocks in the manufacture of goods for coating roads and sidewalks. In all respects, paving slabs comply with the requirements of regulatory documents for concrete paving slabs.

To obtain products of excellent quality, it is very important when vibropressing to comply with a given water-cement ratio. Goods are molded from rigid mixtures with a low water-cement ratio (W/C = 0.25–0.35). N. A. Rybiev [58, 113] studied the features of the formation of the structure and properties of cement stone during compaction by pressing. It was established that cement stone has the greatest strength and density at an optimal ratio W/C = 0.24. A study of the structure of a cement stone using a scanning electron microscopy showed that, with an optimal ratio W/C, a cement stone is characterized by a dense qualitative structure, which indicates that during the process of pressing all the liquid is transferred to the film state, and the films have a minimum thickness and are continuous. The minimum value of the ratio of strength to the degree of hydration

Table 5.17 Physico-mechanical properties of the paving slabs of the experimental batch

No.	Strength, MPa				Frost resistance, grade	Frost resistance factor	Abrasion, g/cm^2	Water absorption, %
	On compression		on Bending					
	R_{com}	$R^{28}{}_{com}$	R_b	$R^{28}{}_b$				
1	42.4	56.5	7.6	8.9	F300	1.22	0.68	2.35
2	31.4	42.0	5.7	7.2	F250	0.96	0.70	2.77
3	35.9	47.8	7.1	8.5	F250	3.12	0.88	3.05
4	37.5	49.3	4.3	5.6	F20	0.97	0.81	3.88
5	38.8	55.4	5.8	6.9	F20	1.13	0.79	2.15
6	37.7	52.2	4.6	5.4	F20	1.20	0.77	2,18

Note: Compositions 1–3 based on fillers from ash-slag waste, water-cement ratio W/C = 0.295–0.34; compositions 4–6 based on fillers from burnt rocks with plasticizing fine additive, W/C = 0.23–0.25; R_{com} and R_b are the compressive and bending strengths after heat treatment; $R^{28}{}_{com}$ and $R^{28}{}_b$ are the compressive and bending strengths after 28 days of normal hardening

(a) (b)

Fig. 5.14 Microstructure of fine-grained concrete: (**a**) ash concrete; (**b**) burnt rock concrete (magnification (×5000), resolution is 10 μm)

of cement stone with optimal ratio W/C also indicates the highest quality structure. Cement stone with an optimal water-cement ratio is characterized by the highest rate of structure formation and it not only has a denser structure, but is also more resistant to corrosion and temperature effects. According to the obtained indicators of strength, frost resistance, abrasion, corrosion resistance, resistance to alternate wetting and drying, and other properties, these concretes can be recommended for the production of paving slabs, side stones, facing slabs, artificial paving stones, and other small-piece goods.

Concretes, characterized by increased strength, water resistance, corrosion resistance, durability, manufacturability, and economy, in large volumes are required for railway crossing slabs. Such plates are designed to ensure intersection at the same level of roads and railways. Sometimes crossings with a monolithic concrete cover are built. Typically, such coatings are available where there is a lot of traffic through the crossing and little traffic along the railway line. Relocation plates are made of reinforced and high-quality concrete. For these purposes, concrete mixtures with a low water-cement ratio of materials from ash-slag waste and burnt rocks of mine dumps are useful. In this case, fillers (crushed stone, crushed stone and sand mixtures, screenings of crushing rocks, slag mixture) and finely dispersed additives (fly ash, ground burnt rock) can be used. The finely dispersed fly ash in the concrete composition replaces part of the cement while being an active microfiller, improves the grain composition of the fine filler of concrete, helps increase the density of concrete, and increases its water resistance.

Ground burnt rock is also used with concrete in this way. However, the activity of this additive is lower than that of fly ash. Fine ash and burnt rock are the most effective and affordable cementitious additives from technogenic raw materials for binders and concrete. The efficiency factor for introducing finely dispersed additives into the concrete mix in order to replace part of the cement is 0.6–0.7; that is,

1 kg of ash can replace 0.6–0.7 kg of cement; for burnt rock this value is 0.45–0.50. Strength performance factors for ash and burnt rock are approximately the same in both early and later stages of concrete hardening.

By replacing a part of cement with finely dispersed additives, the workability of the concrete mixture improves, which is associated, firstly, with an increase in the volume of binder with the introduction of additives, and secondly, with the plasticizing effect of spherical ash particles with a smooth vitrified surface texture. The plasticizing effect of ground burnt rock is lower than that of ash. This is mainly due to the shape, porosity, and surface roughness of the rock particles. However, in general, the plasticity of the concrete mixture increases, the surface quality improves, and the number of shells and pores decreases. It was found that the use of finely dispersed additives with a specific surface of 500–600 m²/kg allows one to ensure high strength characteristics of concrete. This conclusion is consistent with the author's statement [154] that the optimal dispersion of a mineral additive to cement should be 120–200 m²/kg higher than the dispersion of a clinker mineral (cement). With this use of mineral additives, dense packing of the initial matrix of a mixed binder is realized due to the distribution of finely dispersed particles in coarse-grained voids.

The slag mixture contains up to 30% of fly ash. Slag mix and screenings of crushing rocks are used in concrete instead of sand. Compared to natural sand, they have several advantages. They do not contain clay and silt and other contaminating impurities. According to their grain composition, they can be qualified as a coarse aggregate; the fineness modulus, as a rule, is higher than two. Crushed stone from burnt rocks does not concede in properties and quality to crushed stone from natural rocks. Moreover, the roughness and clean surface of the gravel particles increase its adhesion to cement. Materials from burnt rocks and ash-slag waste are not inert with respect to cement. They actively interact with cement, and participate in the hardening of concrete. As a result, an additional amount of hydrosilicates appears. A denser concrete structure is formed. This provides increased strength and improved other important indicators of concrete. The coefficient of frost resistance in many concrete compositions (both on ash and on rock) is higher than unity. Concrete grades from F200 to F500 were obtained for frost resistance and for water resistance from W10 to W12. For concrete on burnt rock fillers, the water resistance grade is from W8 to W10.

Table 5.18 Bending strength of ash products (bench test)

Composition of concrete, kg/m³						Bending strength, MPa	Cement saving, %
Cement	Fly ash	Sand	Slag	Crushed stone	Water		
415	–	473	–	1263	200	39.5	–
353	150	452	–	1236	200	49.8	15
353	–	–	1788	–	180	45.3	15
332	–	–	473	1206	215	46.4	20
290	150	–	1701	–	195	47.8	20
415	–	–	1726	–	195	42.2	–

Table 5.19 Physico-mechanical properties of concrete for railroad crossing plates

Concrete kind	Strength, MPa				Frost resistance, grade factor	Water absorption, grade	Abrasion, grade	Cement saving, %/kg
	R_{com} 28 days	R_b 28 days	R_{com} 1 year	R_b 1 year				
Ordinary concrete	46.6	5.6	49.8	6.13	F300	W8	G2	–
Ash concrete	56.7	8.82	70.43	10.08	F500	W12	G1	20/93
Burnt rock concrete	55.3	8.33	68.47	9.78	F400	W12	G1	15/70

The increased physico-mechanical properties of concrete are achieved at a lower cement consumption than in concrete without additives. The use of materials from industrial raw materials in concrete mixtures allowed one to reduce cement consumption from 10 to 50% without decreasing the properties and quality of products.

At setup, the goods from ash concrete and burnt rocks of mine dumps were tested. The compositions and strength of goods from ash concrete natural hardening, at the age of 28 days, are given in Table 5.18.

According to the test results of plates with materials from burnt rocks, with the design grade of concrete 300, safety factor for crack resistance is equal to 2.7, and the value of the residual deflection after unloading is 5.5 mm. Table 5.19 shows the indicators of physical and mechanical properties of concrete on materials from burnt rocks and ash-slag waste.

Concrete is a capillary-porous system. The smaller the pore sizes between the hydrosilicates, the higher the strength of concrete and its other characteristics that affect the durability of concrete [9]. The more perfect the structure of concrete, the less likely the accumulation of local destruction zones in concrete with the formation of cracks under various variable loads or climatic influences, for example, during alternate freezing and thawing, exposure of aggressive environments, etc.

Chapter 6
Physicochemical Fundamentals of Hardening the Burnt Rocks of Mine Dumps and Ash-slag Waste in Road Pavements

Abstract This chapter presents physicochemical fundamentals of hardening the burnt rocks and ash-slag waste used in road pavements. The discussion is carried out for different constituents of road construction (subgrade, base, and structural layers of pavement). The effects of fillers from burnt rocks and ash-slag waste on the structure formation of asphalt concrete and concrete are also considered with estimation of ways to attain high strength of such structures. In particular, among the hardening products, neoplasms formed during hydration and hardening of clinker minerals are noted. Moreover, thermal properties of hydration products are studied during hardening of concrete.

Keywords Subgrade hardening · Flaky and needle-like grains · Reinforcing effect · Dust fractions · Organo-mineral composite · Surface roughness of particles · Adhesion · Hydration products · Hydrosilicates

6.1 Subgrade Hardening

A general condition for the use of technogenic raw materials in road construction is the compliance of the strength of the structure constructed from it with the mechanical and physico-mechanical effects that can be expected in each layer of pavement. Long-term service of any structure can be provided that the material of the structure meets the conditions of its operation. To ensure reliable resistance of the structure to increasing loads throughout the entire service life, it is necessary that the energy of structural bonds in the material of the structure is constantly increased. This is possible provided that the design has a self-densifying system. Materials from burnt and re-burnt mine rocks and ash-slag waste can create a monolithic self-compacting system without introducing a binder. This ability of the rocks is due to their chemical and mineral composition, structural features, and size. Materials from burnt

A. A. Lyapin et al., *Improving Road Pavement Characteristics*, Innovation and Discovery in Russian Science and Engineering,
https://doi.org/10.1007/978-3-030-59230-1_6

rocks and ash-slag waste of a certain granulometric composition have the ability over time to cement and harden the structure of the bulk of the embankment, to form a monolith of increased bearing capacity and stability. First of all, this is due to the optimal packing of particles of the used material of a certain size and the construction technology of the structure.

At the beginning of compaction in loose placer, individual grains of rock and gravel are easily distributed and mutually displaced. At the same time, large fractions of crushed stone play the role of a kind of spatial framework. Small particles fill the voids of the frame. Flaky and needlelike grains of burnt rock and slag fillers are, in a certain sense, elements of short reinforcement, and affect the adhesion strength between particles, facilitating their mutual physical interweaving, creating the effect of dispersed reinforcement [86]. The reinforcing effect is manifested in the formation of spatial and compact heteropolar bonds, in an additional increase in the strength of the massif.

Materials from burnt rocks and ash-slag waste are well compacted at optimal humidity. When materials are compacted in the layer, large grains are crushed and partially crushed due to face cleavage at the contact points, which increases the content of dust fractions. The particle size distribution of the system is optimized. Fine grains and fine particles are distributed in coarse-grained intergranular voids. An optimal packing of particles is created, and the structure density is high. Thus, an improvement in the particle size distribution and an increase in the number of contacts in the system are associated with the physical factor of hardening the composition. Moreover, due to mechanical destruction, additional energy centers with free surface energy arise on the surface of the particles. These factors contribute to the manifestation of physical and adsorption interactions and the formation of new complexes. Over time, these complexes create a complex, branched spatial framework, the strength of which is higher than the strength of individual microaggregates. A dense, solid structure of the embankment mass with improved physical and mechanical properties is formed.

6.2 Increase of Strength in the Bases and Structural Layers of Pavement from Technogenic Raw Materials

The basis of obtaining compositions for the base and structural layers of roads is the principle of contact hardening, which consists of the ability of dispersed systems of the optimal ratio of components and grain composition to form a solid monolith at the time of convergence of particles during compaction. The materials studied from ash-slag waste, burnt rocks of mine dumps, and metallurgical slag are inherently inactive, but not entirely inert. The strength of the structural layer is ensured by cementation, depending on the activity of the additives (fly ash, ground burnt rocks, slag). Finely dispersed additives not only exhibit binding abilities, but also improve the particle size distribution of the mixture, contributing to a denser packing of particles.

There is a slow strength growth. At an early stage, under conditions of normal hardening, the action of finely dispersed additives reduces mainly to an improvement in the particle size distribution and an increase in the interaction contacts in the mixture. In the initial period of formation of the structural layer, the physical factor prevails. In hardening the structure of the massif, there is direct contact of particles during compaction (physical factor). The latent physicochemical activity of finely dispersed additives arises when even small doses of cement or lime are introduced. Cement additives play the role of activator of the hidden cementing ability of waste [127]. The active components contained in the waste enhance the hydrolysis of cement. They bind lime formed during the hydrolysis of cement into hydrosilicates of various basicity. The adsorption and physicochemical factor are manifested. The total number of compounds strengthening the structure of the resulting layer increases. Over time, when hardening the structure of the layer, the cementing ability of materials from technogenic raw materials predominates, due to their potential activity. Strength growth is a long process, during which strength is first determined by interaggregate interactions based on intermolecular, complex heteropolar, adsorption coordination, and other bonds. Further hardening of the structure is associated with the formation of cementitious compounds with strong condensation-crystallization bond types [109, 116].

6.3 The Formation of the Structure of the Asphalt Concrete Composition Based on Technogenic Raw Materials

It is possible to obtain asphalt concrete composites with high physical and mechanical properties based on industrial raw materials only creating the optimal structure and ensuring strong adhesion between the organic binder and mineral components. The nature of the bonds in the organo-mineral composite is determined by physical, physicochemical, and mechanical interactions at the interface. The basis of the asphalt concrete composition is mineral particles from burnt rocks of mine dumps and ash-slag waste. The proposition [79] is known that the surface of mineral materials of acidic rocks is practically inactive. However, according to research [164, 165] contrary to the conventional wisdom, the surface of acidic silica-containing mineral materials is not inert with respect to the components of bitumen. Moreover, it is known that the surface of all solid materials contains acidic and basic centers of the Lewis and Bronsted types [53, 137, 138], which are obvious and determine its activity with respect to binders (inorganic and organic). The largest contribution to this interaction will be made by Lewis acidic and basic Bronsted centers.

The studied burnt rocks and ash-slag waste are classified as acid rocks. Due to their origin, ash-slag waste and burnt mine rocks have cracks and other structural defects and active components: metakaolinite, amorphous oxides of iron, aluminum, glass phase, and other reactive products. It affects the surface properties of the waste particles, namely the concentration of active centers. As it is known [77, 157], adsorption products of hydration products of mineral and organic binders occur at such centers. Obviously, for burnt rocks of mine dumps and ash-slag waste, evaluating their activity, it is necessary to take into account structural defects, presence of active components that appeared due to their formation, and Lewis and Bronsted centers (synergistic effect). The surface roughness of particles of technogenic raw materials also plays a positive role. It is higher than that of particles of traditional fillers. Adsorption of organic molecules of bitumen occurs by the donor-acceptor mechanism. Thus, despite the acidic nature, materials from technogenic raw materials actively interact with bitumen, providing strong adhesive contacts between the organic compounds of bitumen and the surface of mineral aggregates. This has a positive effect on the physico-mechanical properties of asphalt concrete and its durability.

The studies on the use of non-ore materials (fillers) from burnt rocks and ash-slag waste as part of asphalt concrete show its good adhesion to bitumen, which contradicts the idea of the interaction of bitumen with acidic rocks. The endurance of asphalt concrete largely depends on the adhesion of mineral aggregates to the surface. The adhesion of binder to the surface of mineral burnt rock and ash-slag particles is good, and the strength in the contact zone is a serious obstacle to the formation of microcracks. The structure of organo-mineral asphalt concrete based on materials from burnt rocks of mine dumps and ash-slag waste is characterized by a mixed type of bond. When mineral particles interact with bitumen, various forms of organic mineral and heteropolar complexes are formed. Complex heteropolar organic mineral compounds are formed upon the joint manifestation of ionic and covalent or ionic and coordination bonds. Between the complexes, a cyclic structure is formed, closed by coordination bonds.

Asphalt concrete based on materials from burnt rocks and ash-slag waste is characterized by high values in strength, frost resistance, and water resistance and low values in water saturation. This can be explained, first of all, by a good adhesion of the active grains of the filler with an organic binder. The bitumen film becomes denser and less susceptible to aging from external influences than the bitumen film on the surface of a traditional aggregate. Secondly, this is due to the lack of clay and clogging impurities, and the optimal grain composition. Thirdly, the presence of active components in the composition of burnt rocks and ash-slag waste, as well as their ability to exhibit pozzolanic activity, affects the additional hardening of the material structure, which manifests itself over time after reaching design strength.

6.4 The Effect of Materials from Burnt Rocks and Ash-slag Waste on the Structure Formation of Concrete

The theoretical basis of the methods for producing concrete compositions is the principles of physicochemical mechanics of disperse systems [109, 116]. The basis of these principles is the creation of conditions for directional structure formation and composition formation of a given structure and properties. To increase the number of neoplasms in colloidal, and then in crystalline form, it is necessary to enrich the phase composition with substances capable of reacting with the active part of cement. Such substances are the active components of ash and burnt rock. A study of the hardening processes of concrete with additives showed that both clinker components of cement and mineral additives of ash or burning rock take part in the structure formation. This point of view is also confirmed by the results of studies of other authors [77, 157, 158].

The hardening processes of binders with hydraulic additives such as ashes and burnt rocks have not yet been fully studied and are distinguished by the complexity and variety of the ongoing interactions of the mixture of components with water [7, 8, 42, 155], in which there are clinker parts, and also amorphized clay substance, glass phase, silica in the form of pozzolan, and other active constituents. Fine dispersed hydraulic additives (fly ash, ground burnt rock), showing the properties of pozzolans, in the concrete mix are microfiller and activator of cement. With their presence, the volume of cement paste increases. Pozzolanic activity of additives is manifested in the binding of free lime, which is formed during the hydration of cement by the active components of fly ash or burnt rock.

A decrease in the concentration of calcium hydroxide in the system accelerates the hydrolysis of clinker minerals, thereby increasing the utilization of cement. The number of hydrated compounds is increasing. The formation of hydrates occurs mainly due to surface reactions. Fine particles of additives are distributed between coarse particles, increasing the specific surface of the system and demonstrating the effect of fine powders [88]. On the developed surface of the particles, as on a substrate, the nuclei of new phases of cementing substances get deposited and then there is their growth in the pore space of a cement stone. So, for example, after 6 months, an artificial mixture of cement with ash may contain 1.8–2.2 times less free calcium hydroxide [96] than a purely cement mixture.

Due to the adsorption and molecular and capillary forces on the surface of the filler, a dense and strong contact zone is formed between the filler and the cement stone. Compaction and hardening of the structure of the material occur. This contributes to a denser structure of the hardening composition. In such a system, internal stresses and cracking are reduced. As a result of the intensifying action of additives of fly ash or burnt rock on cement and its hydration products, additional structural bonds are formed in the particle contact zone. The resulting additional compounds fill the voids and pores of the contact zone. As a result, the porous structure of the artificial stone changes: the number of small pores and capillaries increases. Indirectly, a decrease in the porosity of the structural composition with ash or burnt rock can be judged by the amount of water absorption. For such systems, it was 2–3%. Adsorbed water in such micropores will be in a film state. It has an oriented structure and altered properties [46, 47]. The increase in the number of small pores positively affects the strength, frost resistance and water resistance, and deformability of concrete, which is confirmed experimentally.

According to the theory of structure formation at the first stage, a skeleton of a coagulation-type structure is formed. The strength of this structure is ensured by intermolecular forces. The layers of water weaken these forces. These structures are of weak strength. They are thixotropic and restore their mechanical properties, broken by shaking. As hydration processes deepen, the volume of hydrated compounds increases. This leads the system to cramped conditions (supersaturation), and also leads to compaction of the gel phase and contact of particles. There is an intergrowth between the crystals of the neoplasms, their growth, and transformation into condensation-crystallization structures of increased strength and durability. The high strength of such structures is due to strong ionic and covalent bonds. The structure becomes more homogeneous and denser. The introduction of additives of ash and burnt rock, with pozzolanic activity, increases the efficiency of cement by binding free calcium hydroxide to low-basic calcium hydrosilicates, such as C_3SH_2, C_3SH (A), C_4H_{13}, and hydroaluminates C_3AH_6 and C_4AH_{13}.

Figures 6.1, 6.2, and 6.3 present comparative data of X-ray phase and thermal analyzers obtained on the examples of ash concrete and ordinary concrete. Among the hardening products, neoplasms formed during hydration and hardening of clinker minerals are noted. Clinker components and active components of mineral additives of fly ash and ground burnt mine rock take part in the structure formation. It should be noted that in the presence of additives, the phase composition of neoplasms does not change significantly compared to the control one; the difference is only in their quantitative content. However, there is an increase in the number of neoplasms. Fine particles of ash or burnt rock are located between the individual grains of the clinker part of the hardening cement. Under these conditions, favorable conditions are created for its hydration. Thus, free lime released during hydration of tricalcium calcium silicate is first adsorbed and then chemically bounded by the active components of mineral additives. A co-binder cement stone (cement plus additive) after 6 months contains almost half as much free calcium hydroxide as cement without additives. This is a long process; therefore, systems containing hydraulic additives are characterized by an increase in strength after reaching the design. The low strength of the artificial stone with the addition of fly ash or burnt rock at the beginning of the formation of the structure is compensated by higher strength in the later stages of hardening. The final strength can exceed the design 2–3 times.

Under the same conditions of hardening, fly ash, due to the presence of a glass phase, shows a more active participation in structure formation in comparison with a burnt rock. This can be judged by the greater intensity of the neoplasm lines during X-ray studies. This is consistent with data on adsorption activity, characterized by a change in optical density upon adsorption by ash and burnt rock of methylene blue dye. The adsorption activity for the ash is 0.031; for the rock, it is equal to 0.022 [23].

The composition of the resulting compounds indicates that almost all active and reactive ash components are involved in the formation of hydrated compounds. The study of the authors of [5] found that the crystals of low-basic hydrosilicates of the C-S-H (I) type possess the highest strength of hydrated compounds.

All components of the concrete mix, both finely dispersed additives and fillers from burnt rocks and slags, participate in the formation of the concrete structure. Fillers from this technogenic raw material are not inert components. The hardening of the material in the presence of active fillers occurs as a result of chemical and physicochemical interactions, as well as due to the reinforcing effect.

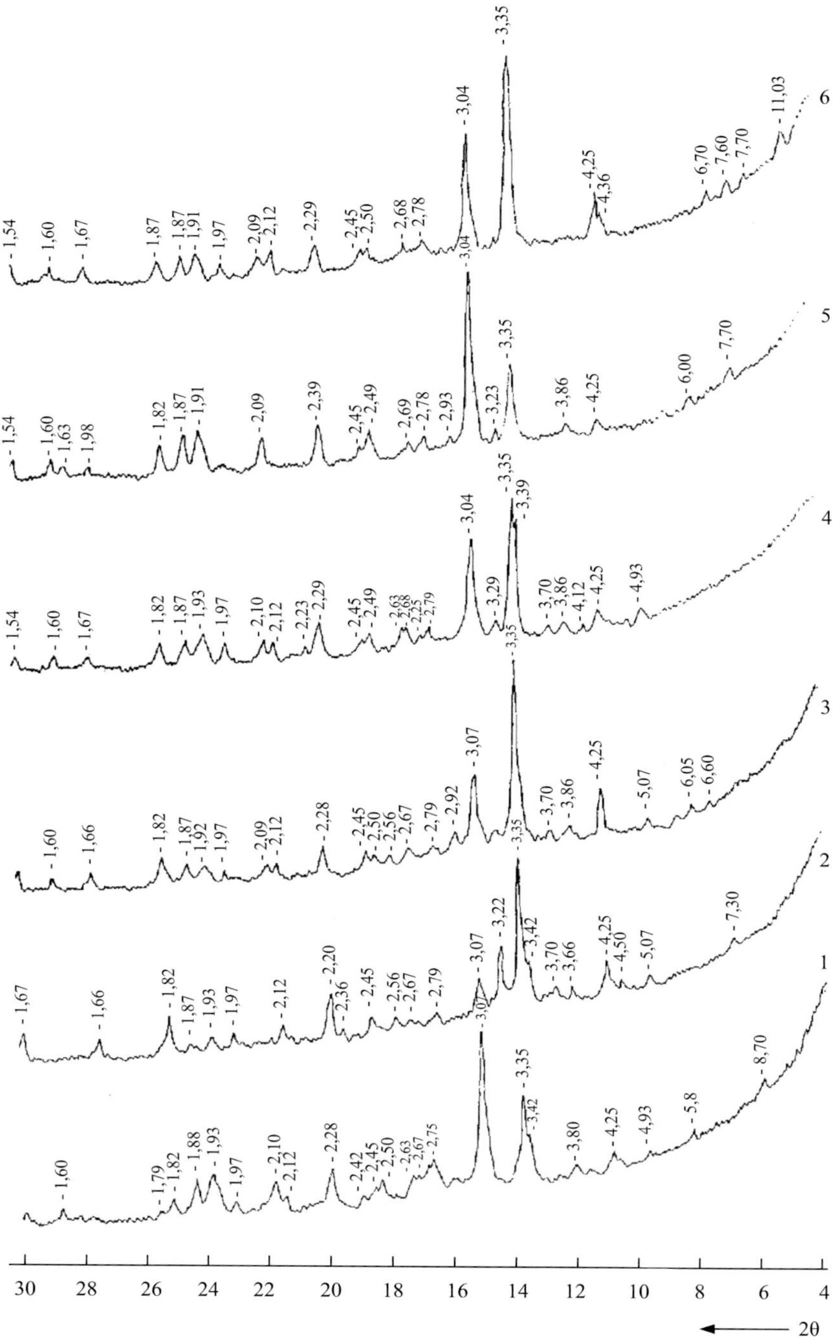

Fig. 6.1 Radiographs of hydration products during hardening of concrete: 1, 2, 3—natural hardening; 4, 5, 6—heat and humidity treatment; 1, 4—control; 2, 5—ash addition with 200 kg/m^3; 3, 6—ash addition with 250 kg/m^3

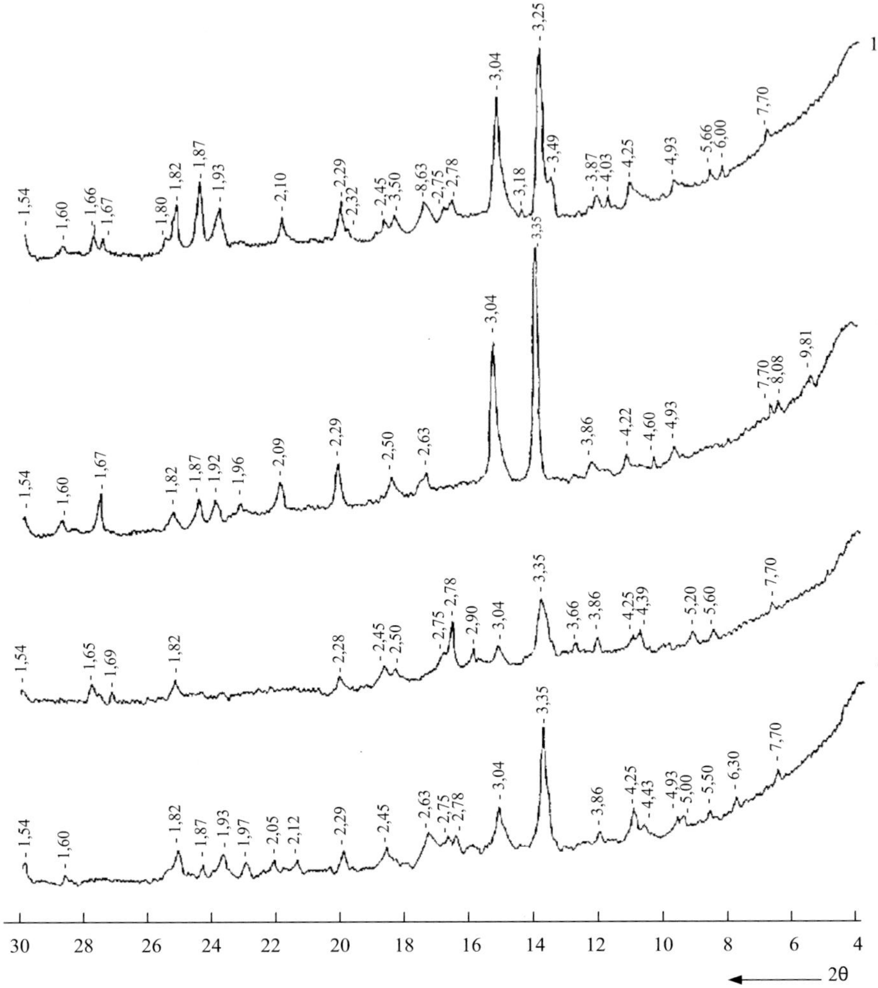

Fig. 6.2 Radiographs of hydration products during hardening of concrete: 7—control, strength grade 150; 8—ash addition with 420 kg/m^3; 9—control, strength grade 300; 10—ash addition with 200 kg/m^3

In thermograms (see Fig. 6.3), samples of ordinary concrete have basically the same effects as ash concrete with different ash contents. The main endothermic effects occur at a temperature of 140–180 °C. The first is associated with the removal of adsorption-bound water and partial dehydration of calcium hydrosulfoaluminate and ettringite. The second is associated with the decomposition of calcite formed due to the carbonization of calcium hydroxide. The endothermic effect at 160 °C, characteristic of many samples of ash concrete, is associated with the dehydration of hydroaluminates and calcium hydrosilicates. The endothermic effect at 480–500 °C is associated with the removal of chemically bound water from calcium hydroxide.

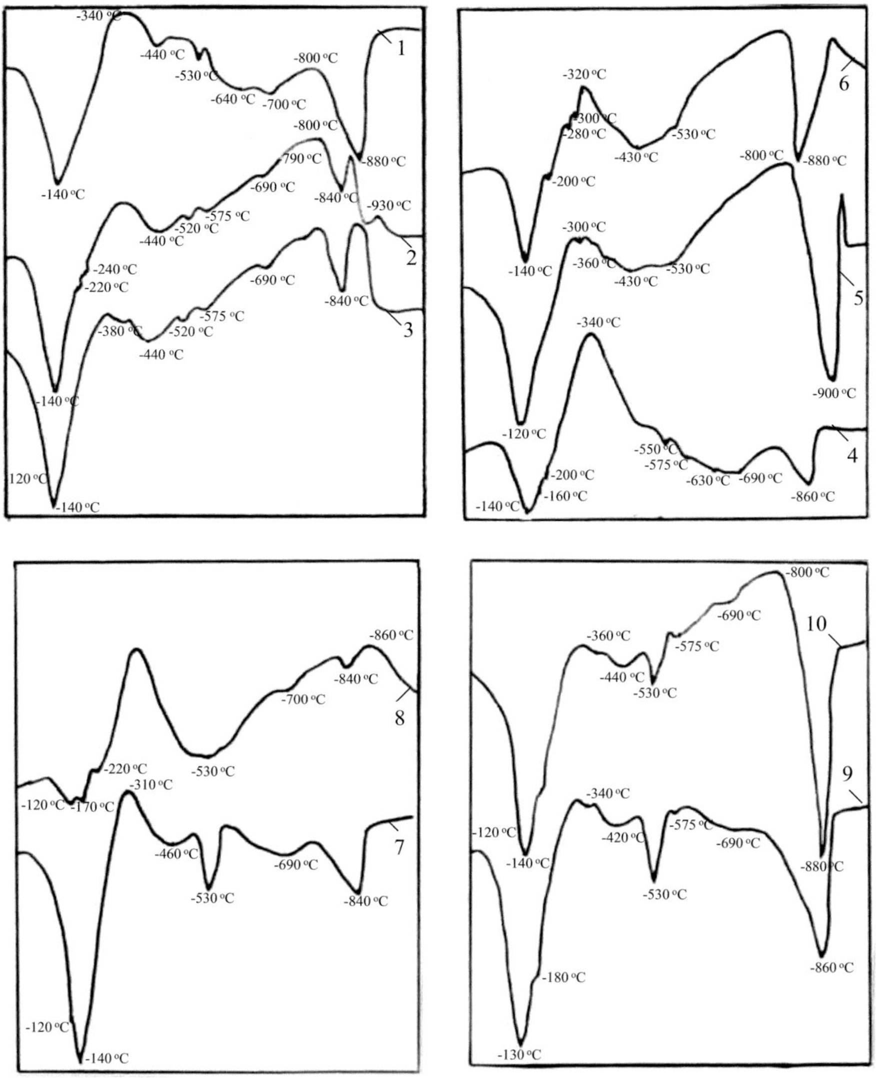

Fig. 6.3 Thermograms of hydration products during hardening of concrete: the designations are the same as in Figs. 6.1 and 6.2

With an increase in hardening time, this effect decreases, which indicates a more complete binding of lime to neoplasms and an increase in the degree of cement hydration. At 530 °C, an effect characteristic of Portland cement occurs. In samples of ordinary concrete, the value of the latter is very significant. In samples with ash, this effect decreases and is practically absent for some compositions. Endothermic effects at a temperature of 400 °C indicate the presence of $C_2SH(A)$ hydrosilicates. Moreover, highly basic calcium hydrosilicates $C_2SH(C)$ and C_3SH_2

appear, characterized by a group of endothermic effects at 690 °C, 710 °C, and 750 °C. The endothermic effect at 560 °C can be associated with a decomposition of hydrogarnets. The exothermic effect at 890–910 °C indicates the presence of weakly crystallized low-basic hydrosilicates of the tobermorite group.

The characteristic features of concrete containing ash-slag waste and burnt rocks of mine dumps should be noted. During heat and moisture treatment, the reactivity of the active components of ash-slag waste and burnt rocks is significantly increased, and the number of hydrated cementitious compounds increases. The greatest increase in strength is observed during long-term storage of such concrete in conditions of high humidity. High physical and mechanical properties of concrete with additives of ash-slag waste and burnt rocks are confirmed by the practice of using products from such a concrete.

Chapter 7
On the Efficiency of Using the Burnt Rocks of Mine Dumps and Ash-slag Waste in Road Constructions

Abstract This chapter is devoted to ecological and economic assessments of the efficiency of using the burnt rocks of mine dumps and ash-slag waste in road constructions. These estimations are based on the prevented damage to environment and also assessment of the efficiency of technogenic waste use as an industrial subsector being an element of the regional economic complex system directed to complete satisfaction of the social economic needs of the region.

Keywords Ecological and economic assessment · Utilizing burnt mine rocks · Prevented damage · Fly ash · Crushed stone · Energy state

7.1 Ecological and Economic Assessment of the Utilization of the Burnt Rocks of Mine Dumps

The main task of ecological and economic assessment of waste management is to draw up a balance of costs aimed at the implementation of a technical solution, taking into account possible returns, expressed in the cost of the damage prevented.

A comparative analysis of the efficiency of the two directions of using burnt mine rocks is carried out below: (i) construction of an access road and (ii) laying of a developed underground space.

When considering the ecological feasibility of disposing of burnt mine rocks, it is necessary to analyze all types of impacts of the planned activity at all its stages and give them a quantitative assessment. A comprehensive indicator reflecting the level of negative impact of anthropogenic activities on the state of the environment is the damage expressed in monetary form.

The amount of damage prevented can be defined as the difference between environmental damage to the environment before and after the implementation of the evaluated technical solution.

A. A. Lyapin et al., *Improving Road Pavement Characteristics*, Innovation and Discovery in Russian Science and Engineering,
https://doi.org/10.1007/978-3-030-59230-1_7

The basic option is the construction of a road using natural raw materials: crushed stone obtained by crushing rocks mined in quarry at the distance of 30 km from the construction of the road. The production of crushed stone in the base case is associated with the following anthropogenic impacts:

(i) Lands are taken away under the quarry.
(ii) Extraction of natural raw materials, and violation of the landscape of the earth's surface.
(iii) Crushing, loading, unloading of materials, air pollution with dust and exhaust gases of working equipment in the quarry and at the crushing and screening site.
(iv) Delivery of crushed stone to the construction site, air pollution by automobile exhaust gases.

In the basic version, the underground space is not laid; as a result, the area under the mine is not used due to disturbance of the earth's surface. In options I and II, crushed stone obtained from burnt mine rocks, which are coal mining waste, is used. At receipt of burnt rock crushed stone, the same types of anthropogenic impact on nature are taken into account as in the production of crushed stone from natural raw materials. The annual volume of processed burnt rocks is 100,000 tons. The distance from the waste dump to the construction site is 30 km, and from the dump to the laying site, it is equal to 20 km.

To conduct a comparative analysis of the two directions of utilizing burnt mine rocks, a quantitative assessment of the environmental damage at various stages of the handling of natural raw materials and waste was performed (see Table 7.1). The calculation of environmental damage was carried out in accordance with the basic regulatory and methodological documents currently in force [67, 139, 140].

The total values of the prevented damage under option I (E_1) and option II (E_2) are given below:

$$E_1 = 67.568 + 72.8 + 99.78 + 282.846 + 4.87 + 15.755 + 1232.28 - 377.124 - 3.683 - 1232.28 = 162.812 \text{ rubles}$$

$$E_2 = 67.568 + 72.8 + 966.964 - 2.361 - 179.146 - 850.261 = 75.564 \text{ rubles}$$

As the calculations showed, the values of the prevented damage in both cases are of positive importance, which indicates the environmental nature of the estimated areas of waste utilization. Comparison of the results allows one to conclude that the disposal of burnt mine rocks during the construction of road foundations gives a far greater environmental and economic effect and is more preferable than the laying of the developed underground space with these wastes.

Table 7.1 Damage to environment in basic version and as a result of utilizing burnt mine rocks

Option	Stage of waste generation	Type of environmental impact	Damage, thousands of rubles/year
Basic	Storage of mine rock in the dump	Taking away territory	67.568
		Environmental pollution (air, water, soil)	72.800
	Quarry mining and production of crushed stone	Quarry territory rejection	99.786
		Dusting during the extraction and processing of natural raw materials and handling	282.846
		Blasting air pollution	4.870
		Emissions from the extraction and processing of natural raw materials	15.755
	Transportation of crushed stone to the construction site of vehicle	Vehicle exhaust	1232.280
	Developed space is not bookmarked	Territory rejection	966.964
Option I (road construction) 1232,280	Dump development and production of crushed stone from burnt mine rocks	Dusting during the development of a dump, processing burnt mine rocks, and loading and unloading	377.124
		Emissions from the development of dump and the processing of burnt mine rocks	3.683
	Transportation of crushed stone from burnt mine rocks to the construction site of the road	Automobile exhausts	1232.280
Option II (laying of the developed underground space)	Dump development and crushing of burnt mine rocks	Exhausts during the development of the dump and crushing burnt mine rocks	2.361
		Dusting during the development of dump, crushing of burnt mine rocks, and loading and unloading	179.146
	Transportation of crushed burnt mine rocks to the site of laying works	Exhausts of vehicles	850.261

7.2 Efficiency of Application of Fly Ash and Burnt Ash in Building Materials

The study examines the most objective approach to assessing the efficiency of industrial waste use, based on the industry as a sub-sector of an element of the regional economic complex system. The purpose of the functioning of such a system is the most complete satisfaction of the social economic needs of the region.

The building materials industry is one of the most resource-intensive sectors of the national economics. Concrete and reinforced concrete are the building materials, without which it is almost impossible to build a single capital construction. These materials are the main consumers of cement and fillers. It is possible to reduce the consumption of these components and the cost of concrete by using products from technogenic wastes that are not inferior in quality to natural raw materials and do not worsen the properties and quality of concrete. Such products from industrial raw materials include materials from ash-slag waste and burnt rocks of mine dumps.

This section shows the efficiency of the use of fly ash generated at Novocherkasskaya TTP and fillers from burnt rocks of mine dumps, available in the Rostov region (Russia), as concrete components. Numerous scientific and technical studies and production results have proved that ash-slag waste and rocks of mine dumps can be considered as an important source of raw materials for the production of cement, concrete, crushed stone, sand, and other building goods. Long-term tests of products obtained during the processing of burnt mine rocks by the enterprises "Tract May," "Trans USA," "TempDorStroy," "Technostroy," "Anfo," and others give reason to conclude that fractionated gravel, crushed stone mixtures, screening crushing, and sand from crushing screenings, according to all technical characteristics, meet the requirements of regulatory documents that apply to similar products from natural raw materials.

As it is known [105], in heavy concrete the use of fly ash helps to improve the quality of the multicomponent matrix and, ultimately, improve the construction and technical properties of the finished good. Most often, ash is used in concrete as an active mineral additive, microfiller, and plasticizer and is introduced into the composition of concrete immediately at the time of preparation of concrete. In this case, ash is used as a multifunctional additive, replacing part of the cement and sand.

The use of ash-slag waste and burnt rocks in concrete gives not only a saving in raw materials, but also a reduction in the cost of production. As an example, data of an elementary calculation (only the cost of raw materials is taken into account) of the efficiency of the use of such materials in concrete are given (see Table 7.2).

The calculations show the average cost of materials. There are various options for replacing traditional raw materials with industrial ones.

Option I. Dry ash (fly ash) is introduced instead of part of the cement, namely cement consumption is reduced by 30% and sand by 10%.

Option II. Dry ash replaces 30% cement and 10% sand and a burnt rock filler (fractioned gravel from crushed burnt rocks) is introduced.

Table 7.2 Comparative data on the cost of concrete of the basic composition with fly ash and materials from burnt rocks

Materials	Cost, rubles	Need for raw materials per month, tons	Cost of raw materials, thousands of rubles									
			I		II		III		IV		V	
			Basic	Proposed	Basic	Proposed	Basic	Proposed	Basic	Proposed	Basic	Proposed
Cement	5500	500	2750.0	1925.0	2750.0	1925.0	2750.0	2475.0	2750.0	2557.5	2750.0	1925.0
Fly ash	760	240	–	182.4	–	182.4	–	–	–	–	–	182.4
Crushed stone natural	350	1400	–	–	490.0	490.0	–	–	–	–	–	–
Sand natural	180	850	153.0	–	153.0	137.7	153.0	162.0	153.0	159.4	153.0	–
Crushed stone from burnt rocks	210	1400	–	–	–	–	–	294.0	–	294.0	–	294.0
Screening crushing burnt rocks	80	850	–	–	–	–	–	–	–	–	–	68.0
Cost, thousands of rubles			2903.0	2245.1	3393.0	2735.1	3393.0	2931.0	3393.0	–	3393.0	–
Efficiency, thousands of rubles (per month)			657.9		657.9		461.97		382.1		923.6	
Efficiency, thousands of rubles (per year)			7894.8		7894.8		5543.64		4585.2		11083.2	

Option III. Burnt rock crushed stone is introduced, cement consumption is reduced by 10%, and sand consumption is increased by 5.9%.

Option IV. Burnt rock crushed stone is introduced, cement consumption is reduced by 7%, and sand consumption is increased by 4.2%.

Option V. Burnt rock crushed stone and screening crushing of burnt rocks are introduced instead of natural sand; dry ash is introduced instead of part of the cement, reducing its consumption by 30%.

The magnitude of the economic effect is greater, the closer the slag or mine dumps are located to the object of consumption of these materials.

The most objective approach to assess the efficiency of industrial waste use was proposed by the authors of [16]. It is based on the consideration of the industry or sub-industry as an element of the system of the national economic complex of the Rostov region (Russia). The purpose of the functioning of such a system is the most complete satisfaction of the social economic needs of the region. With this approach, the total economic effect of the use of industrial waste in construction is determined by the formula

$$E_{\text{eff}} = E_{\text{mp}} \pm E_{\text{reg}} \pm E_{\text{c}},$$

where E_{mp} is the effect obtained in the sphere of material production of the region; E_{reg} is the effect obtained in the social sphere of the region; and E_{c} is the effect obtained in construction from the use of industrial waste.

The effect obtained in the sphere of material production in the region is the volume of profit of industrial enterprises from the sale of production waste:

$$E_{\text{mp}} = E_{\text{al}} + E_{\text{md}} + E_{\text{p}},$$

where E_{al} is the effect obtained as a result of the release of agricultural lands and a reduction in payment for them; E_{ss} is the effect obtained in the social sphere of the region; E_{md} is the effect obtained from reducing the cost of maintaining dumps; and E_{p} is the effect of reduced payment for limited and excess waste.

The area of dumps reaches tens of square kilometers. The cost of 1 hectare of land has a steady upward trend. Moreover, unsold waste significantly increases the cost of production manufactured by industrial enterprises. This is due to the costs of their storage, maintenance, transportation, etc. The introduction of fees for subsoil and land significantly increases the interest of enterprises in the sale of waste.

Great importance has been given to the effect obtained in the social sphere when using industrial waste (E_{reg}). It consists of the effect of improving the environment (E_{ie}) and the effect of including the released land in the national economy of the region (E_{ner}):

$$E_{\text{reg}} = E_{\text{ie}} + E_{\text{ner}}$$

The region has a direct interest in waste utilization; as the area of arable land increases, the levels of pollution of water bodies, soil, and atmosphere decrease; the quality of agricultural products increases; and the terrain landscape deteriorates.

Using the methodology [99], the effect obtained as a result of environmental improvement can be estimated by the formula

$$E_{ie} = E_{lc} + E_{pa} - E_{st} K_{inv},$$

where E_{lc} is the effect obtained by improving the living conditions of the population; E_{pa} is the increase in profit in agriculture from the reduction of environmental pollution; E_{st} is the industry standard efficiency ratio; and K_{inv} is the amount of capital investment in the implementation of environmental measures.

The effect obtained by improving the living conditions of the population consists of a reduction in the loss of standard products during illnesses of workers due to environmental degradation, a reduction in social insurance fund payments for the period of incapacity for work due to an unfavorable environmental situation and healthcare costs for the treatment of such diseases, and increase in labor productivity.

Thus, if the cumulative effect $E_{eff} > 0$, but $E_c < 0$, then to stimulate the use of waste by the enterprises, it is necessary to compensate for their losses. Additional funding for the organization of waste utilization can be allocated from environmental funds, formed by payments from enterprises. The implementation of subsidies from these funds will allow one to link the improvement of the ecological situation in the region with the stimulation of their rational use in enterprises.

7.3 Concluding Remarks

Long-term laboratory and analytical studies, technological and experimental-industrial tests, and practical experience have proved the possibility of using ash-slag waste and burnt mine dumps in road construction. The composition, properties, and condition of these technogenic raw materials provide the opportunity to replace traditional raw materials. However, their use is connected with certain difficulties. One of them is the heterogeneity of the composition and properties of these materials. In this regard, there is a need for thorough preparation of technogenic raw materials, which consists of bringing their properties to the requirements of regulatory documents of a particular industry of their application. There are a number of areas requiring additional research, such as (i) a detailed study of the energy state of the active components of the waste; (ii) optimization of the amount of replaced natural raw materials depending on the activity and properties of the waste; (iii) development of normative documentation aimed at expanding the areas of application of these technogenic raw materials; and (iv) accumulation of data on long-term deformation and strength characteristics, and durability of structures made of such materials. The result of solving such problems will be more advanced technologies for the use of ash-slag waste and burnt mine dumps, with increase in their disposal.

To evaluate ash-slag waste and burnt rocks as technological raw materials, an individual approach is required for each ash and mine dump. The prospects for the use of a specific technogenic raw material depends on the proximity of its properties to indicators that characterize the properties of similar traditional raw materials. Evaluation of technogenic raw materials includes the following steps:

(i) Engineering-geological survey of dumps to obtain the necessary information about the nature of the occurrence of materials, distribution of humidity, density on the depth of the dump, and degree of uniformity
(ii) Assessment of indicators of physical and mechanical properties of ash-slag waste, burnt rocks of mine dumps, and materials from them, determined in the laboratory
(iii) Determination of the possibility of using the studied technogenic raw materials for the construction of the subgrade, foundations, and structural layers of pavement
(iv) Development of a design project and recommendations on the technology of its construction
(v) The implementation of a feasibility study on the appropriateness of using certain materials in specific road structures

The classification of ash-slag waste and the preparation of burnt rocks of mine dumps for use require certain costs. However, the development of natural raw materials is even more expensive [39]. Materials from burnt rocks and ash-slag waste are lower in cost but not inferior in quality, and in some respect superior to similar samples from traditional natural raw materials. Materials from ash-slag waste and burnt rocks of mine dumps will be able to displace similar products from natural raw materials. It is quite possible, but not immediately, since new products still need to win their place in the market of goods. At the same time, one must be active in the competition when it is moving towards the buyer. The unpreparedness of potential consumers of ash-slag waste and mine dump rocks for this activity can become a restricting factor when replacing natural materials by technogenic raw materials.

Part II
Finite-element Modeling of Road Constructions on Soils

Chapter 8
Dynamic Modeling of Solid Mass on Soil Base

Abstract This chapter considers dynamic modeling of solid mass on soil base. Features of finite-element modeling soil massif are discussed on the base optimal selection of representative volume with damping belts. As the test task, it is considered lattice of partitioning a three-layer half-plane into finite elements, with the introduction of damping belts and taking into account a coastal slope. The dynamic stress-strain states of these structures are studied for different geometrical parameters of the layers and their mechanical characteristics. Finally, comparison of finite-element calculations is performed with analytical solutions, obtained by using integral method of harmonic analysis based on the application of the Fourier transform in time.

Keywords Dynamic modeling · Finite-element modeling · Solid mass on soil base · Representative volume · Stress-strain state · Monoharmonic effects · Analytical models · Damping belts · Three-layer structure

8.1 Dynamics of "Solid Mass-Soil" System

The methodology for solving model problems by FEM includes three stages. The first stage is determined by the development and implementation of the FEM calculation model, which allows us to correctly calculate the characteristics of the dynamic stress-strain state of the soil base of a given structure at given loads (stress functions $q_{ky}(x)$) on its surface. The second stage is based on the implementation of the calculation model of the solid mass, at the stresses (equal to the stresses on the surface of the soil massif) given along the boundary of the contact of the solid mass foundation with the soil. The third stage is determined by finding the uncertain coefficients for the stress functions from the condition that the displacements of the object and the soil massif at the contact area are equal.

A. A. Lyapin et al., *Improving Road Pavement Characteristics*, Innovation and Discovery in Russian Science and Engineering,
https://doi.org/10.1007/978-3-030-59230-1_8

8.2 Features of Finite-Element Modeling Soil Massif

By developing a calculation model of a soil massif (in a plane or spatial setting), based on the finite element method and including a coastal slope as an element, the main issue is the correct choice of a representative limited volume of a semi-limited structure. By formulating and solving a plane problem, this problem is technically solved somewhat easier; therefore, the main attention is paid to namely such a statement of the problem. At the same time, the results obtained in the development of a plane model are an important original data for creating a spatial model. The correctness of choice of the size of the representative volume is checked by comparing the main wave characteristics in a given part of the structure, obtained for the layered half-space without breaking the structure using analytical methods [4, 69, 73, 74, 161], and the results of calculations using the FEM model.

Let us discuss in details these issues for the case of a plane model. To obtain the criteria for choosing the representative volume sizes, number, and characteristics (width, viscosity) of damping belts, we consider a model of homogeneous ($y \leq 0$) or layered half-space ($y \leq H$) (Fig. 8.1). It is important to determine the area in which it is necessary to obtain reliable information about the characteristics of the dynamic stress-strain state of the medium. This area must include the zone in which the object is located (it is highlighted in the figures by a double dash-dot line).

The criterion for the initial choice of the size of the representative volume for a homogeneous half-space is the time, T, during which the fastest, namely the longitudinal, wave reaches the boundary of the representative volume and, after reflection from it, attains the boundary of the region in which it is necessary to obtain reliable characteristics of the dynamic stress-strain state.

Let us consider the time interval $t \in [0, T]$. The dimensions A and B of the representative volume of the region depend on the wave propagation velocities in the layer materials. Since the propagation velocity of longitudinal waves $V_{pj} = \sqrt{\lambda_j + 2\mu_j / \rho_j}$

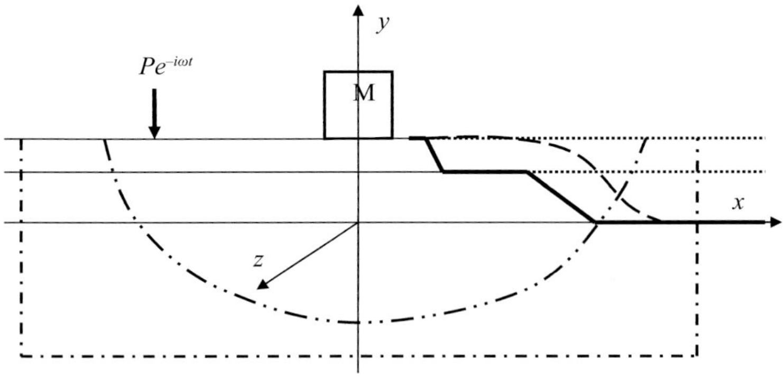

Fig. 8.1 Considered model: layered half-space boundary; • – • – • – representative volume boundary; • — • — • — • — the boundary of the area in which it is necessary to calculate the characteristics of stress-strain state

in the materials of the layers is higher than the propagation velocity of transverse waves $V_{sj} = \sqrt{\mu_j / \rho_j}$, the required dimensions of the representative volume are determined approximately through the largest of them (max{V_{pj}} = V_p) as follows:

$$\begin{cases} 2A - R \geq V_p T \\ 2B - R \geq V_p T \end{cases} \tag{8.1}$$

where R is the distance from the place of dynamic impact to the boundary of the region in which it is required to calculate the characteristics of the dynamic stress-strain state of the structure (Fig. 8.2).

The exact value of time T can be determined by using a numerical experiment. On the plots of amplitude-time characteristic (the dependence of amplitude on time) at any point of the representative volume, one can clearly distinguish the passage of a direct wave from the source of oscillations and the arrival of a wave reflected from the boundary of the representative volume (a fictitious wave that does not exist in a half-bounded structure). Thus, at each point of the region, one can clearly distinguish the time range in which it is possible to correctly carry out the calculation (the period of time before the arrival of the first reflected wave). The initial dimensions of the representative volume of the region can be selected by formula (8.1) and a given value of T.

After the initial selection of the sizes of the representative volume (for example, for a homogeneous half-space, see parameters A and B in Fig. 8.2), we divide this region into finite elements, followed by calculating the characteristics of the dynamic stress-strain state inside the region in which calculations are necessary.

At the same time, several types of partitioning the region into finite elements are necessarily checked. Each subsequent option is characterized by a decrease in the maximum size of the finite element (at this stage only the convergence and accuracy of the calculated data are checked, for example, by using displacements). The accuracy criterion is the comparison of wave fields from nonstationary impact into the area highlighted by a double-dashed line. By this, the time period within which the calculated amplitude-time characteristic of the displacements does not contain waves reflected from the boundary of the representative volume is also determined.

The result of the first stage allows one to determine the maximum size of the finite element necessary to achieve the required accuracy.

At the second stage, we determine the value of the representative volume necessary for the correct calculation of wave characteristics. With this aim, keeping the size of the finite elements that determine the partition of the representative volume, we increase its size (parameters A, B).

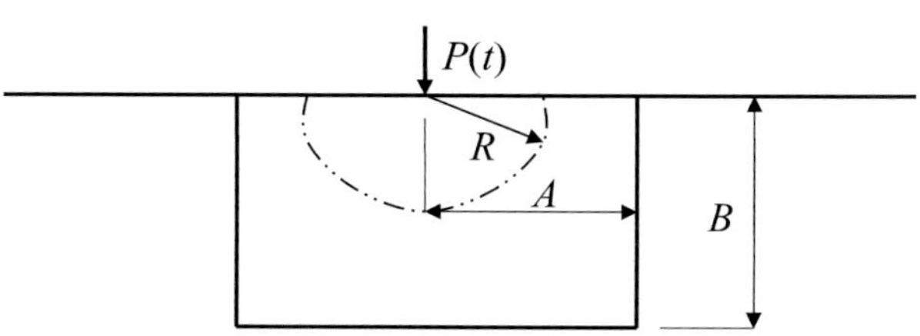

Fig. 8.2 Considered region of calculation of the dynamic stress-strain state

The calculation is performed for monoharmonic effects, followed by obtaining the amplitude-frequency characteristics of the displacements of the surface of the half-space. To control the reliability of the obtained results, we use the results of comparing the calculated data for the FEM model with the frequency response obtained as a result of calculations using an analytical solution that describes the wave field in a homogeneous or multilayer half-space in the form of an improper contour integral in the complex plane [4, 161]. Since the amplitude-frequency characteristic is the most sensitive to the ratios of the geometric and mechanical characteristics of the system, we consider the closeness of the results of calculating the frequency response of the FEM and analytical models as the criterion for optimizing the choice of representative volume sizes.

If we get unsatisfactory comparison results, then the size of the representative volume is increased and the procedure for dividing the area repeated.

In some cases, we find that the dimensions of the representative volume necessary to achieve the required accuracy (especially when calculating for low-frequency harmonics) turn out to be quite large. In this case, the calculation of the characteristics of the stress-strain state of the system elements requires large resources and does not always allow one to achieve the required accuracy. In this case, it is possible to reduce the sizes of the representative volume to be divided into finite elements by introducing two or three damping belts (see Fig. 8.3). Their elements have the same mechanical characteristics as those located in the same layer, but the viscosity increases with distance to the periphery. Damping belts are also divided into finite elements (one element across the width of the belt).

The introduction of damping belts leads to a decrease in the selected sizes of the representative volume of the region, since the intensities of the waves reflected from the boundaries of the representative volume are significantly reduced due to losses when passing through the damping belts. With the introduction of damping belts with the correct choice of viscosity in each of them, one can minimize the energy of waves reflecting from fictitious faces. However, with a strong change in viscosity, reflected waves can appear at their boundaries. To reduce this effect, it is necessary to gradually increase the viscosity of the subsequent damping belt compared to the previous one. The efficiency of the introduction of damping belts and the choice of their characteristics are determined by the results of a numerical experiment.

For a spatial model, the size of the three-dimensional representative volume is selected in each of the three directions determined by the same criteria as for plane problems. Moreover, it is almost always advisable to use the introduction of damping

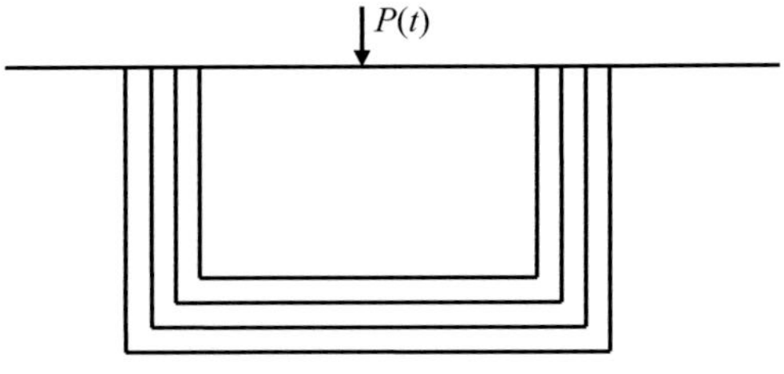

Fig. 8.3 Damping belt layout

belts around the perimeter of the representative volume. This is determined by the fact that in the three-dimensional region the number of finite elements (FEs) in the partition and nodes is significantly larger than in the plane one.

8.3 Finite-Element Model of System: Test Tasks

As a result of choosing the sizes of the representative volume of the region, introducing damping belts and a lattice for partitioning structural elements into FEs, we obtain FEM models of a multilayer half-space and a structure containing a slope. As an example, Fig. 8.4 shows a partition of a three-layer half-plane into FEs with the introduction of damping belts.

Figure 8.5 presents an example of partitioning a three-layer structure, including a coastal slope, into FEs. This structure differs from the layered half-plane in that part of the layers is actually "cut out" in the half-plane (in accordance with Fig. 8.1). The location of the vibration-generating loading is shown by the arrow.

Testing of these models included all the steps described above in Sect. 8.2. Moreover, we used the results of calculations based on analytical representations of the solution to the problem for a multilayer half-space based on an integral representation of the form [4, 91, 161]:

$$\vec{U}(x,y,z,t)=\frac{1}{2\pi\mu}\int_{\Gamma}\int_{\Gamma}\vec{\vec{\mathrm{K}}}(y,\alpha,\beta)\overline{\overline{\overline{P}}}(\alpha,\beta)\exp(-i[\alpha x+\beta z])d\alpha d\beta\exp(-i\omega t)$$

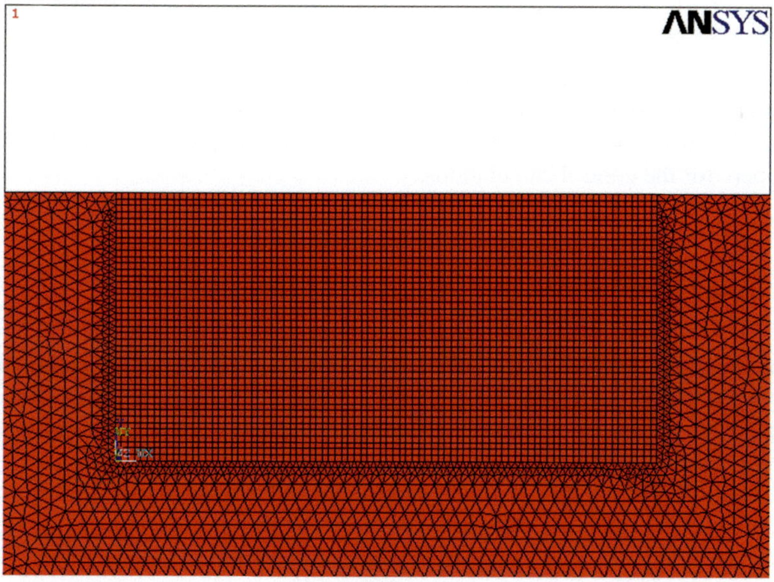

Fig. 8.4 Lattice of partitioning a three-layer half-plane into FEs, with the introduction of damping belts

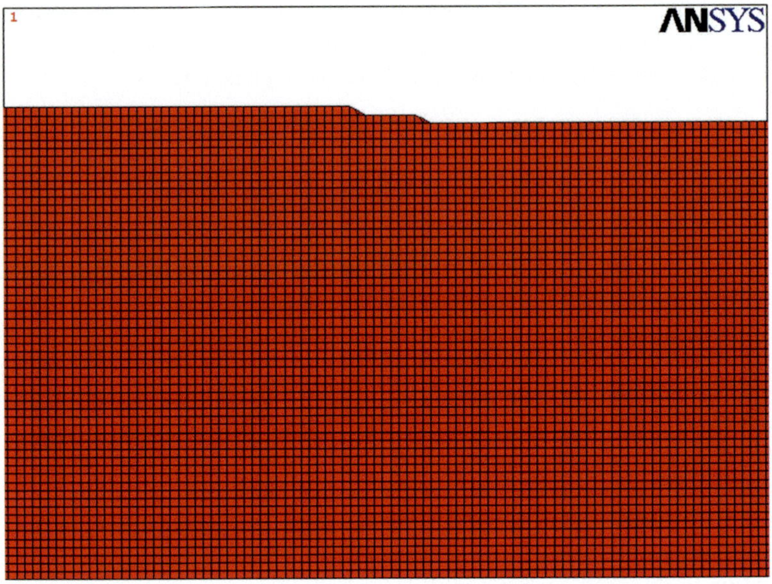

Fig. 8.5 Lattice of the partition of a three-layer structure into FEs, including the coastal slope

Here, the third-order matrix function $\vec{\vec{\mathrm{K}}}(y,\alpha,\beta)$ has a rathy cumbersome structure; the algorithm for obtaining it is described in detail in [91]; $\overline{\overline{P}}(\alpha,\beta)=\int_{-\infty}^{+\infty}\int_{-\infty}^{+\infty}P(x,z)\exp(i[\alpha x+\beta z])dxdz$ (here $P(x, z)$ is the load distribution law acting in the region Ω on the surface of the structure and generating steady harmonic oscillations with frequency ω). This algorithm was implemented by the computer software Strac developed in the Research Institute of Mechanics and Applied Mathematics, Southern Federal University, and kindly provided by the developers for the control calculations.

8.4 Test Calculations and Analysis of Results

By implementing the FEM, test calculations of model problems for various structures were carried out using the ANSYS software. The material properties of a homogeneous half-plane were determined by the following parameters: modulus of elasticity, $E_2 = 8 \times 10^8$ N/m^2;

density, $\rho_2 = 1050$ kg/m^3; Poisson's ratio $\nu_2 = 0.29$; longitudinal wave propagation velocity $V_{p2} = 1000$ m/s; and transverse wave propagation velocity $V_{s2} = 548$ m/s. The selected region Θ, in which it is necessary to perform the calculations, has the shape presented in Fig. 8.6 with sizes of 300×150 m^2.

The estimated time during which waves reflected from the fictitious boundary of the representative volume of the wave reach the boundaries of the region Θ: $T \geq \frac{h}{V_s} = \frac{150}{548} = 0.27\,\text{s}$.

Minimum preliminary sizes of reprvesentation volume: $A = H > V_pT = 1000 \times 0.27 = 270$ m.

In accordance with the above algorithm, to verify the accuracy of the calculations, we also consider the increased size of the representative volume for three cases: (i) $A = H = 300$ m; (ii) $A = H = 360$ m; and (iii) $A = H = 400$ m. An example of dividing a representative volume into rectangular finite elements is shown in Fig. 8.7.

The dimensions of FEs during testing decreased, until the required practical accuracy of calculations was achieved.

The dynamic effect on the system is specified in the form of a nonstationary force, which varies linearly in time (see Fig. 8.8).

After selecting the required FE sizes, it is possible to calculate all the characteristics of the dynamic stress-strain state in the region. The most informative are the plots of the dependence of the displacement amplitude on time, that is, the amplitude-time characteristic of displacements. As an example, Fig. 8.9 demonstrates these plots in the time interval $t \in [0, 1]$ for two points: A (50, 0); B (150, 0).

In the plots, three sections can be distinguished. The first section is the same for all cases and determines the direct field of the surface source of oscillations, characterized by maximum perturbations.

The second section of the plots is characterized by the background value of the displacement amplitudes at the level of calculation error. The duration of this part depends on the sizes of representative volume and is determined by the time during which the oscillations reach the nearest boundary of the representative volume, reflect from it, and attain the observation point.

The third part of the amplitude-time characteristic is determined by the appearance at the observation point of reflected waves that do not exist in the half-plane. Therefore, from the moment of their appearance, the calculated data do not correspond to the solution of the original problem.

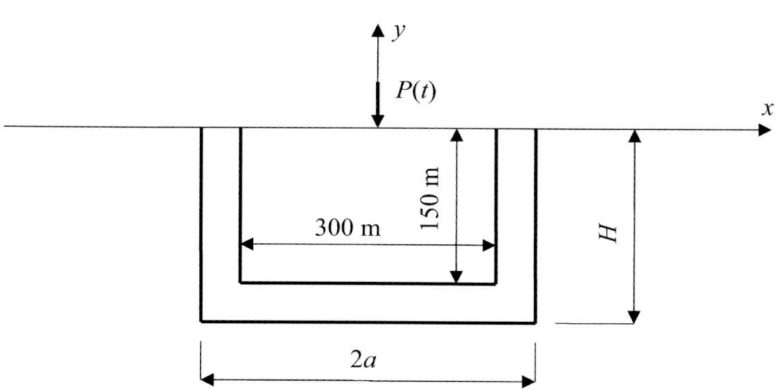

Fig. 8.6 Model structure with the shape and sizes of region Θ, for which calculations in the dynamic problem were performed

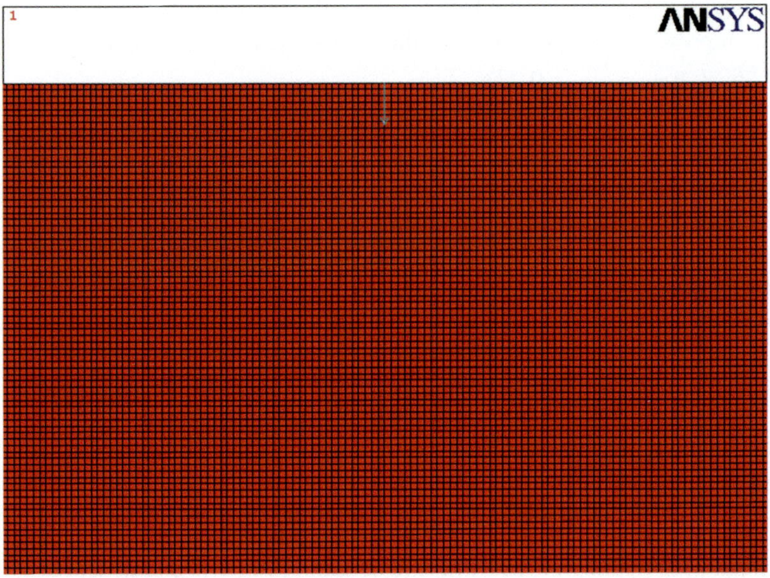

Fig. 8.7 Half-plane partition lattice into finite elements

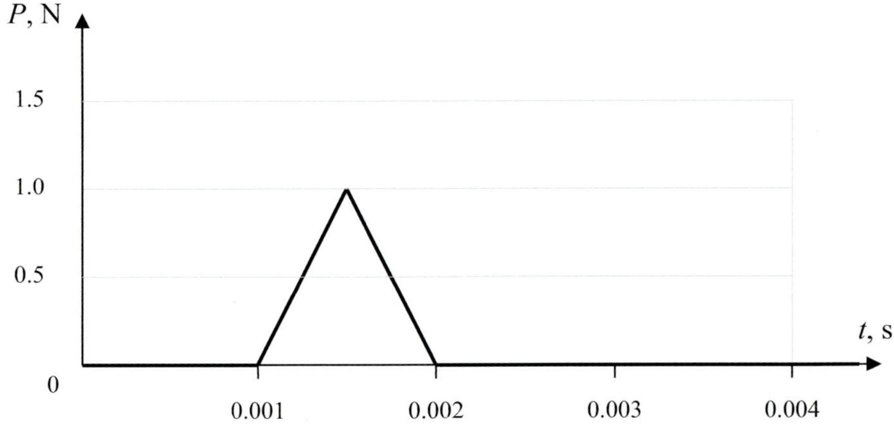

Fig. 8.8 Dependence of load vs. time

Figure 8.10 presents characteristic plots of stresses over time at point A, which have the same sections as the plots of displacements. Only the amplitudes of the stresses corresponding to the reflected waves (the third sections of the plots) are significantly smaller than for displacements (this effect is not significant).

In the plots of changes in the displacements and stresses during time at point B (see Figs. 8.11 and 8.12), the same patterns are observed.

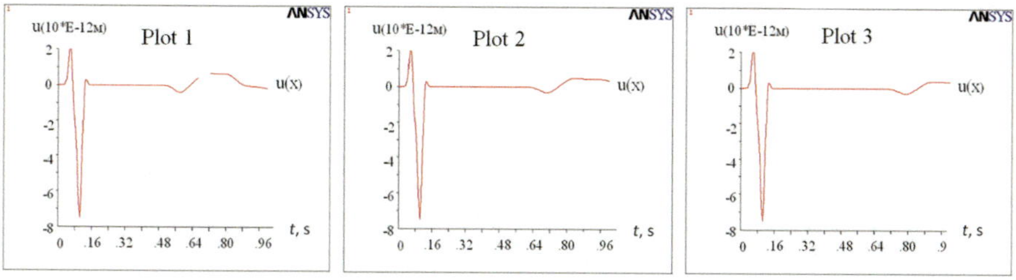

Fig. 8.9 Plots of changes in displacements at point A (50, 0) in time for three dimensions of the region: (i) 600×300 m^2; (ii) 720×360 m^2; and (iii) 800×400 m^2

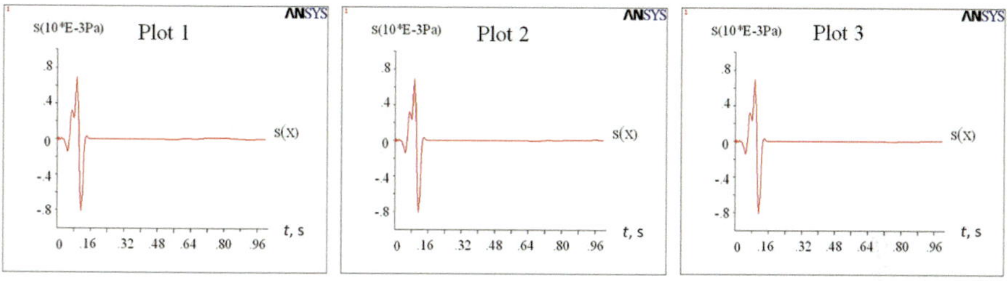

Fig. 8.10 Plots of changes in stresses at point A (50, 0) in time for three dimensions of the region: (i) 600×300 m^2; (ii) 720×360 m^2; and (iii) 800×400 m^2

The difference is because point B is located at a distance from the area of application of the load and the reflected waves appear in it much earlier and have higher amplitude. Obviously, as the size of the representative volume increases, the reflected waves in the region Θ become less energetic, with a simultaneous increase in the time interval in which they do not distort the distribution of stresses and displacements. This means that the analysis should consider only the section of the plots on which there is no influence of reflected waves.

From the plots of changes in stresses and displacements during time at a given point for various sizes of the representative volume, it is obvious that only in the second and third cases, see Figs. 8.11 and 8.12 (for dimensions of the representative volume 720×360 m^2 and 800×400 m^2), we can distinguish the section of the plots on which there is no influence of reflected waves. This means that in order to obtain reliable information on the characteristics of the dynamic stress-strain state of the structure in the area with a diameter of 150 m, we must select the size of the representative volume as $a = H > V_pT = 1000 \times 0.36 = 360$ m.

In order to reduce the size of the representative volume, test calculations were carried out with the introduction of three damping belts (see Fig. 8.13).

The dimensions of the representative volume, including three damping layers, correspond to 600×300 m^2. The lattice for dividing the representative volume into rectangular FEs is shown in Fig. 8.4.

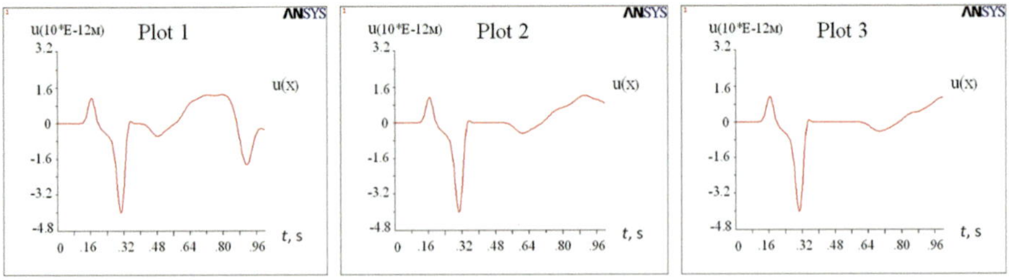

Fig. 8.11 Plots of changes in displacements at point B (150, 0) in time for three dimensions of the region: (i) 600×300 m^2; (ii) 720×360 m^2; and (iii) 800×400 m^2

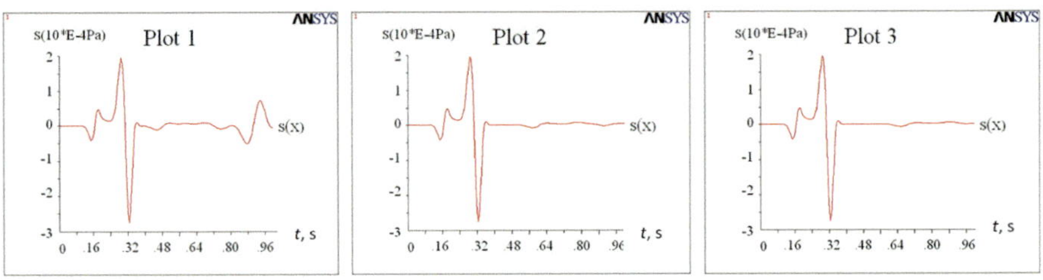

Fig. 8.12 Plots of changes in stresses at point B (150, 0) in time for three dimensions of the region: (i) 600×300 m^2; (ii) 720×360 m^2; and (iii) 800×400 m^2

The mechanical characteristics of the material of the damping belts correspond to the material of the medium; the viscosity is set to increase from the inner layer to the next in a ratio of 2:5:10. Test calculations were carried out for a model problem with various thicknesses of damping belts.

In order to obtain reliable information about the characteristics of the dynamic stress-strain state of the structure in the area with a diameter of 150 m, as examples Figs. 8.14 and 8.15 show plots of the amplitude-time characteristic of displacements and stresses in the time interval $t \in [0, 1]$ for point B (150, 0).

The plots show that only with the introduction of damping belts the influence of reflected waves significantly decreases. As the thickness and number of these belt zones increase, the reflected waves decrease in amplitude and actually approach the level of calculation error.

For comparison, test calculations of model problems for a homogeneous half-plane containing a slope with a height of 5 m and a width of 10 m were performed (see Fig. 8.16). The material properties and the region in which it is necessary to carry out the calculation are chosen as in the case of a model problem without a slope. The partition lattice of the structure, including the coastal slope, into finite elements is shown in Fig. 8.5.

The most informative for determining the influence of slope on displacement are estimates of the plots of dependence of the amplitude of displacements on time (amplitude-time characteristics of oscillations) and dependences of displacements in the vicinity of the slope at different times. As examples, Figs. 8.17, 8.18, 8.19,

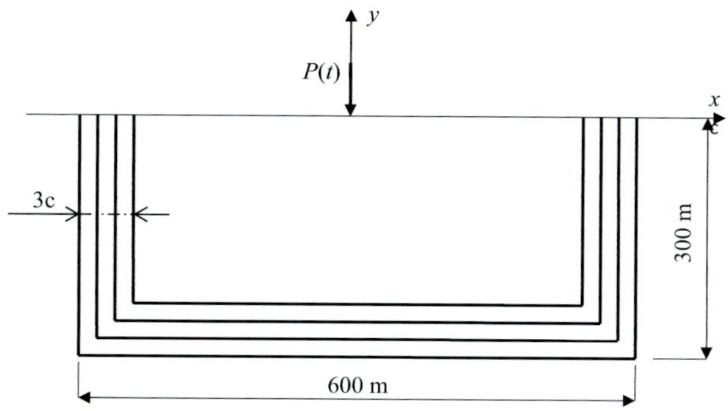

Fig. 8.13 Schematic structure of the model with the introduction of damping belts

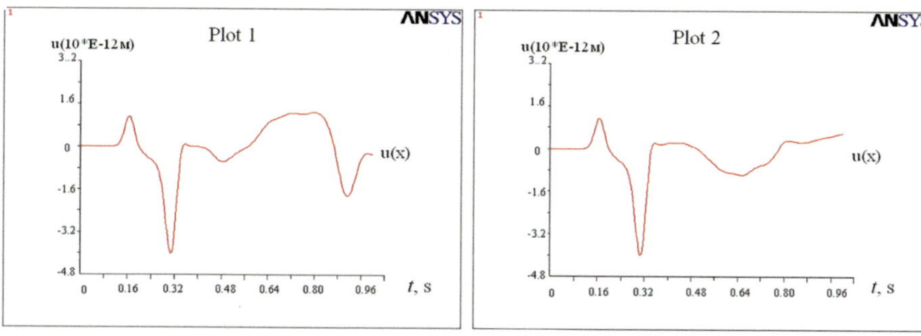

Fig. 8.14 Plots of changes in displacements at point *B* in time for two cases: (1) region without damping belts and (2) introduction of damping belts with a thickness of 20 m

8.20, 8.21, 8.22, and 8.23 show distributions of the amplitude-time characteristics of displacements in the time interval $t \in [0, 1]$ for some points and dependences of the displacements of soil at $t = 0.12$ s in various cases.

The distributions of displacement amplitudes in the studied area at different times are presented in Figs. 8.17, 8.18, and 8.19. Figure 8.17 shows that in the absence of a slope in the structure, there is a symmetry of the displacement distribution relative to the area of action of the load.

However, this pattern changes when a slope appears (see Figs. 8.18 and 8.19). For this case, from the moment the leading wave front reaches the slope boundary, a violation of the symmetry of the displacement distribution law is observed. When the slope is located on the right side of the loading (Fig. 8.16, case 1) in the distribution in Fig. 8.18 there is a slight increase in the amplitudes of displacements in the area of the slope and in front of it.

In the case when the slope is located to the left of the loading (Fig. 8.16, case 2) in the distribution in Fig. 8.19, it is seen that the amplitudes of the displacements in the vicinity of the slope are slightly reduced.

Figures 8.20 and 8.21 present plots of the amplitude-time characteristics of displacements, respectively, for points *A*' (−50, 0) and *A* (50, 0) (see Fig. 8.16, case 1).

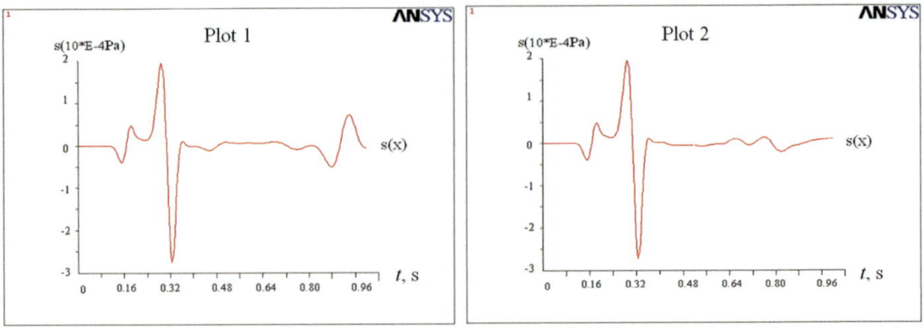

Fig. 8.15 Plots of changes in stresses at point *B* in time for two cases: (1) region without damping belts and (2) introduction of damping belts with a thickness of 20 m

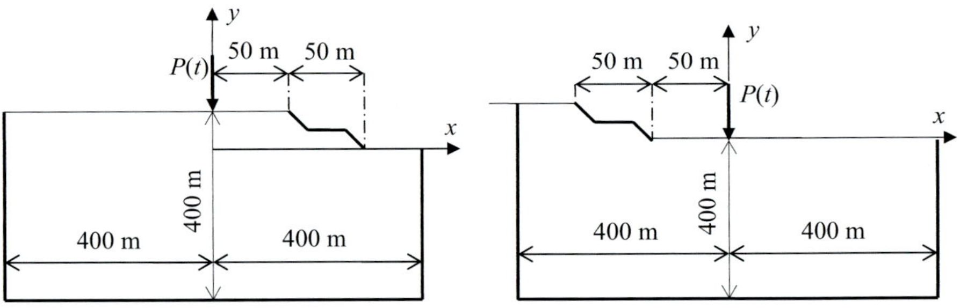

Fig. 8.16 Schematics of the model problem with slopes in different sides from the source of dynamic impact

In the plots given, it is seen that the existence of a slope determines the appearance of waves reflected from it. In this case, to the left of the slope, the reflected waves are manifested by a relatively weak growth of displacement amplitudes. To the right of the slope, the amplitude of the propagating waves decreases markedly.

Similarly, when the slope is located to the left of the oscillation source (Fig. 8.16, case 2), in Figs. 8.22 and 8.23, waves are also reflected from the slope on the right side and the wave amplitude decreases after it passes through the slope.

So, the presented results reflect the general laws of the distribution of displacement amplitudes obtained by using fairly extensive numerical experiments.

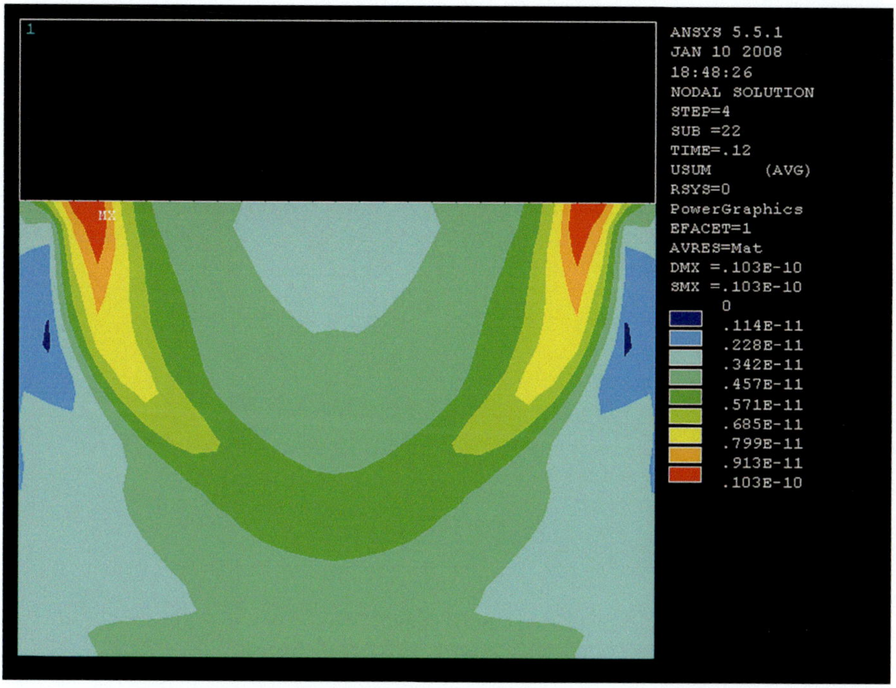

Fig. 8.17 Distribution of displacements in the structure without a slope at $t = 0.12$ s

8.5 Comparison of Finite-Element Calculations with Analytical Solutions

Analytical methods for constructing solutions to the problems of excitation and propagation of oscillations in a multilayer half-space are based on the formulation of problems on forced harmonic oscillations. Real processes in the "solid mass-slope" system are usually associated with nonstationary influences of various types. To use the solutions of harmonic problems to describe nonstationary processes, well-developed methods of harmonic analysis in integral or discrete form are used [40, 131]. It is advisable to use these methods when comparing calculated and experimental data.

The integral method of harmonic analysis is based on the application of the Fourier transform in time to all characteristics [15, 131]. For example, the direct Fourier transform in time to the displacement vector has the form

$$\bar{\vec{u}}(x,y,z,\omega) = \int_0^\infty \vec{u}(x,y,z,t)\exp(-i\omega t)\,\mathrm{dt}$$

As a result, a solution of the problem in the time domain (nonstationary) reduces to solving problems in the frequency domain, followed by returning to the time domain by applying the inverse Fourier transform. For example, the transition from $\bar{\vec{u}}(x,y,z,\omega)$ to $\vec{u}(x,y,z,t)$ is performed according to the formula

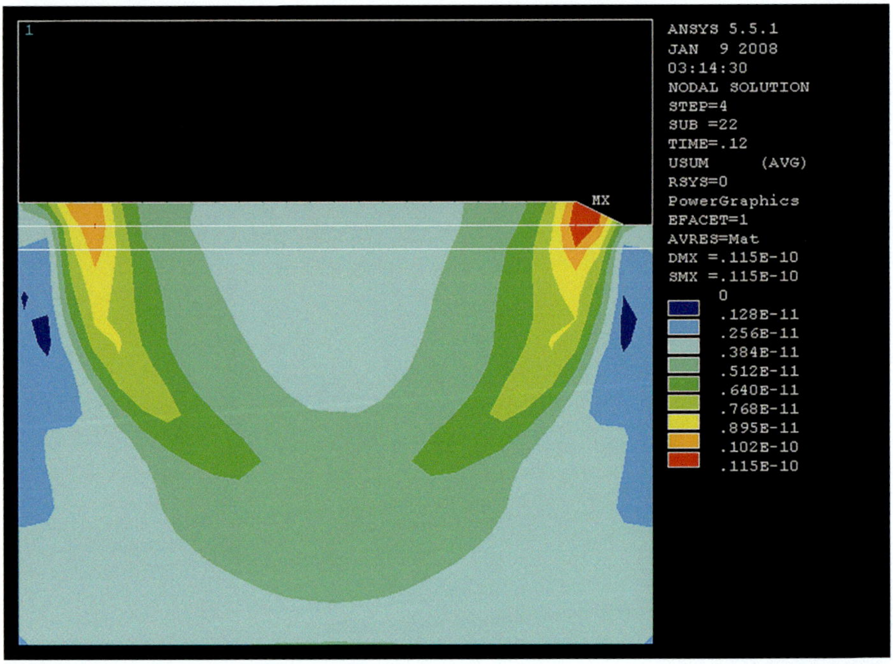

Fig. 8.18 Distribution of displacements in the structure in Fig. 8.16 (case 1) at $t = 0.12$ s

$$\vec{u}(x,y,z,t) = \int_0^{\infty} \bar{\vec{u}}(x,y,z,\omega)\exp(i\omega t)\mathrm{d}t \tag{8.2}$$

It should be noted that formula (8.2) is obtained by applying the method of harmonic analysis in integral form.

The method of harmonic analysis in discrete form is based on the expansion of all functions in a Fourier series in time. Next, solutions are constructed for monoharmonic oscillations with frequencies determined by the Fourier series. The description of the nonstationary process is obtained in the form of a Fourier series with coefficients determined from the solutions of the corresponding harmonic problems.

Based on the solution of the harmonic problem for loading with a unit amplitude, it is possible to construct an intrinsic amplitude-frequency characteristic (AFC) of the displacement of a given point. This frequency response will correspond to the frequency spectrum of oscillations of a given point under nonstationary loading by a force impulse described by the delta function $\delta(t)$ [94]. By using the method of harmonic analysis in integral form for a linear system, we obtain the frequency response (AFC) of oscillations under the loading by an arbitrary impulse of force by multiplying the eigen AFC of this characteristic on the AFC of a given loading.

By comparing the model characteristics calculated by the FEM, for the nonstationary impact corresponding to Fig. 8.8, with analytic, we can use the method of harmonic analysis. To do this, apply the Fourier transform in time to the obtained

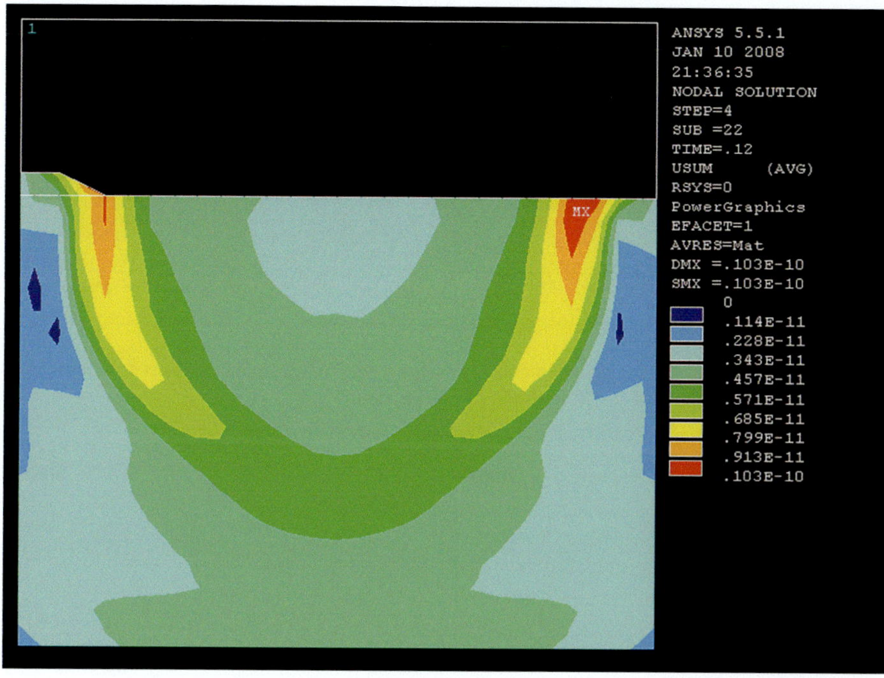

Fig. 8.19 Distribution of displacements in the structure in Fig. 8.16 (case 2) at $t = 0.12$ s

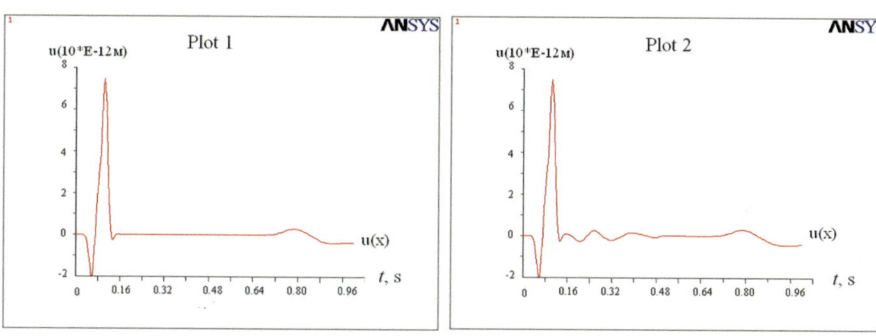

Fig. 8.20 Plots of changes in displacements at point *A*' (−50, 0) over time in various structures: plot 1—structure in Fig. 8.6; plot 2—structure in Fig. 8.16, case 1

amplitude-time characteristic (see, for example, Figs. 8.9 and 8.11 for displacements) and obtain the corresponding AFC, which can be compared with that obtained on the base of calculations by the analytical model. In this case, it should be taken into account that the frequency response, for example, displacement, corresponding to nonstationary loading such as in Fig. 8.8, may have a normalizing factor. To determine this, it is necessary to construct the AFC of the loading $\overline{P}(\omega) = \int_0^\infty P(t)\exp(-i\omega t)\,dt = \int_0^{T_0} P(t)\exp(-i\omega t)\,dt$ (here T_0 is the time at which

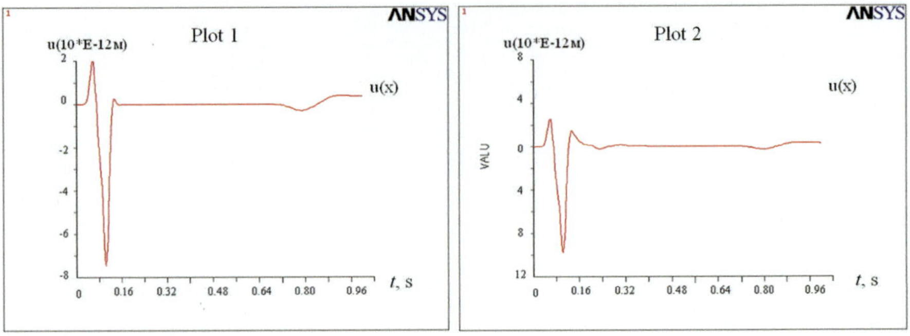

Fig. 8.21 Plots of changes in displacements at point *A*' (50, 0) over time in various structures: plot 1—structure in Fig. 8.6; plot 2—structure in Fig. 8.16, case 1

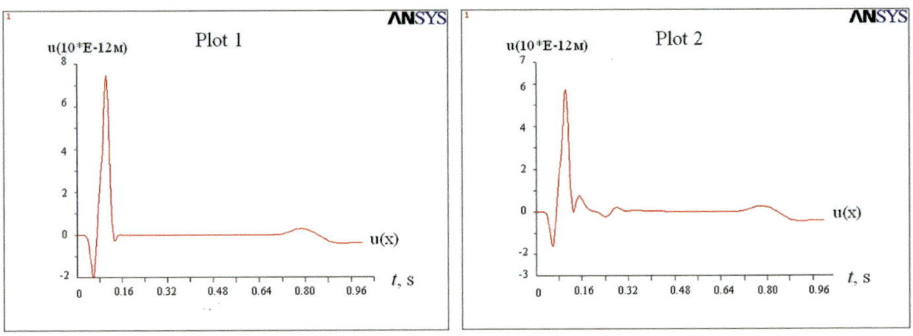

Fig. 8.22 Plots of changes in displacements at point *A*' (−50, 0) over time in various structures: plot 1—structure in Fig. 8.6; plot 2—structure in Fig. 8.16, case 2

the function $P(t)$ vanishes; for Fig. 8.8, we have $T_0 = 0.0011$ s). Figure 8.24 shows an example of the frequency response of the loading, corresponding to Fig. 8.8.

In the low-frequency range, the frequency response plot is close to a constant value ($P_0 = 5 \times 10^{-4}$), and then it decreases with the growth of ω and tends to zero. The eigen frequency response obtained by solving the problem for the delta function $\delta(t)$ corresponds to the impact with a constant spectrum of unit amplitude. Therefore, the normalizing factor for the given case is $P_0^{-1} = 0.2 \times 10^4$:

$$\left|\vec{U}\right|\left(10^{-13}\ \text{m}\right)$$

By calculating the frequency response of the displacement at a given point (for example, for the plot of the amplitude-time characteristic displacement, see Fig. 8.9b), we obtain a plot that has some qualitative differences from analytically obtained (see Fig. 8.25).

These differences are because of the graph of amplitude-time characteristic, according to which the frequency response is plotted, and there is a perturbation associated with the presence of a wave reflected from the boundary of the representative

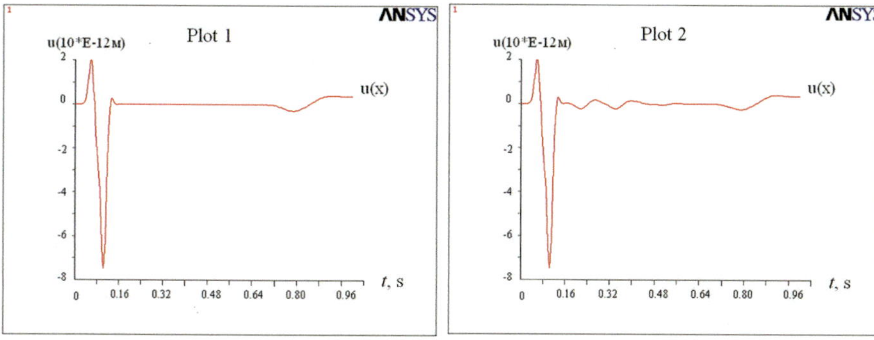

Fig. 8.23 Plots of changes in displacements at point A' (50, 0) over time in various structures: plot 1—structure in Fig. 8.6; plot 2—structure in Fig. 8.16, case 2

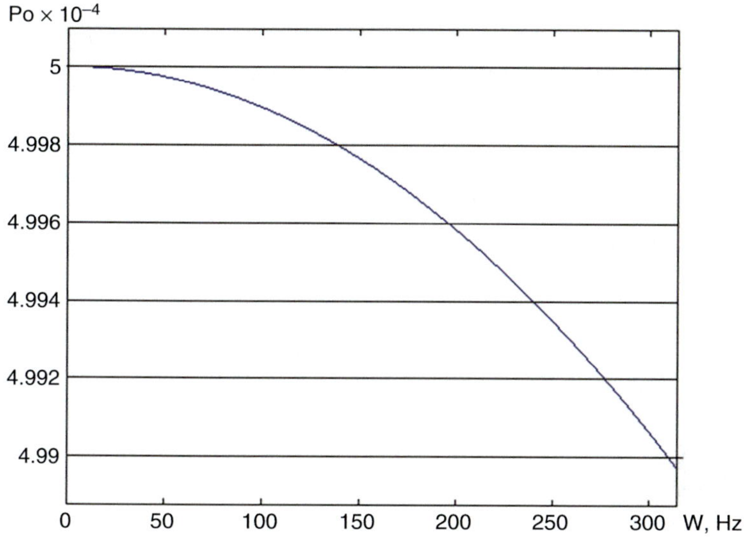

Fig. 8.24 Frequency response corresponding to Fig. 8.8

volume (from time $t_1 = 0.5$ s). Let us introduce filtering of the amplitude-time characteristic by multiplying it by a function $\chi(t)$ equal to unity in the time range corresponding to the passage of the main wave, followed by a continuous decrease and vanishing until the time t_1 of arrival of the reflected wave (see Fig. 8.26). Then as a result of subsequent application of the Fourier transform, the perturbation of the frequency response, associated with the presence of the reflected wave on the amplitude-time characteristic graph, will disappear and the difference in frequency response, obtained by the FEM calculation and calculation by analytical representations, will be almost within the calculation error.

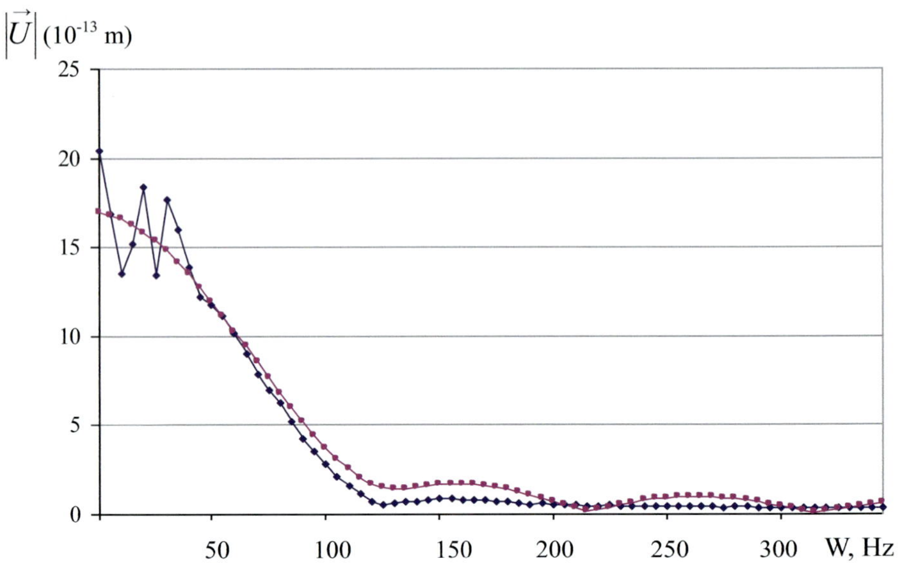

Fig. 8.25 Curve marked with dots is the analytical AFC; the plot marked with squares is the calculation of the frequency response according to the FEM results

Fig. 8.26 Plot of function $\chi(t)$

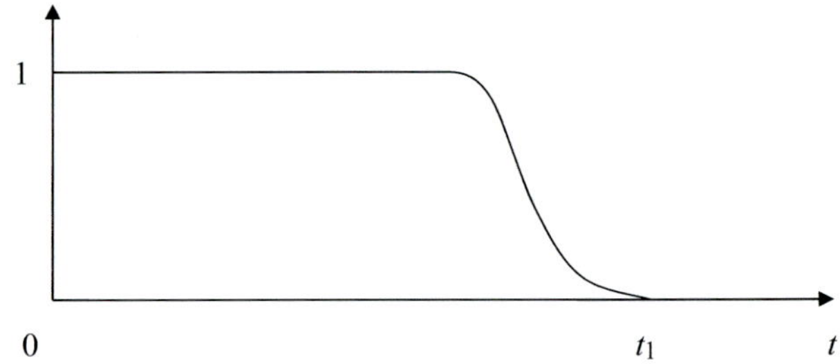

Thus, the results of test calculations confirm the correctness of choosing the sizes of the representative volume and the possibility of using the FE model for calculating the amplitude-time characteristic and frequency response in the required region of the layered half-plane based on the algorithms described above.

Chapter 9
Studying Characteristics of Waves Propagating in Layered Structure with Semi-infinite Layers

Abstract This chapter deals with the base of numerical experiments and the basic laws of the distribution of the amplitude characteristics of oscillations propagating in soils of various structures in the presence and absence of a coastal slope. The consideration distinguishes two cases of the ratio of the stiffnesses of the layers of the investigated structure: (i) the normal structure, when the stiffnesses of the layers increase with the depth of the layered structure, and (ii) the anomalous structure in which the pointed regularity is violated. Two types of anomalous structures with a distribution of layer stiffnesses from top to bottom are considered: (i) "hard–soft–hard" and (ii) "soft–hard–soft." Moreover, these numerical results are obtained taking into account that an oscillation load is located to the left or right of the slope.

Keywords Waves in layered structure · Semi-infinite layers · Hard–soft–hard system · Soft–hard–soft system · Source of oscillations · Slope of layered structure · Horizontal and vertical displacements · Maximum amplitudes

In this chapter, based on a numerical experiment, we study the basic laws of the distribution of the amplitude characteristics of oscillations propagating in soils of various structures in the presence and absence of a coastal slope. As noted above, it is advisable to distinguish two cases of the ratio of the stiffnesses of the layers of the investigated structure. Further, by the structure of the normal structure we mean a layered structure in which the stiffnesses of the layers increase with depth. By anomalous structure we mean a structure in which this regularity is violated. Two types of anomalous structures with a distribution of layer stiffnesses from top to bottom are considered:

(i) "Hard–soft–hard"
(ii) "Soft–hard–soft"

A. A. Lyapin et al., *Improving Road Pavement Characteristics*, Innovation and Discovery in Russian Science and Engineering,
https://doi.org/10.1007/978-3-030-59230-1_9

To identify general regularities, a numerical experiment was performed, the results of which are systematized, and the main patterns of the distribution of oscillations in the medium for each type of its structure are revealed. For this:

(i) Calculations of the distributions of the amplitudes of oscillations in the layered structure (normal and anomalous structures) are performed for the model problem in the FEM implementation.
(ii) A description is given of the main laws governing the distribution of wave fields in a layered soil with a slope in its vicinity for each type of layered structure.
(iii) A comparison is made of the basic laws of wave distribution in a normal and anomalous structure.

The main results illustrating the conclusions are presented in the form of plots and patterns of characteristics.

9.1 Normal Structure (Stiffness of Layers Increases with Depth)

The study of the problem of the characteristics of waves propagating in a layered structure is carried out on the basis of the implementation of the FEM using the ANSYS software package. We consider the problem in a plane setting. As a modeled structure, a slope was chosen, including two semi-infinite layers with inclined ends (an inclination angle of 26.5°) and thicknesses $h_1 = h_2 = 5$ m rigidly fixed with each other and with the underlying half-space (Fig. 9.1). The distance of the stepped surface between the slopes is set: $CD = 30$ m.

For a three-layer structure of a normal structure, the stiffnesses of the layers increase with depth. The mechanical properties of the material are determined by the following basic parameters:

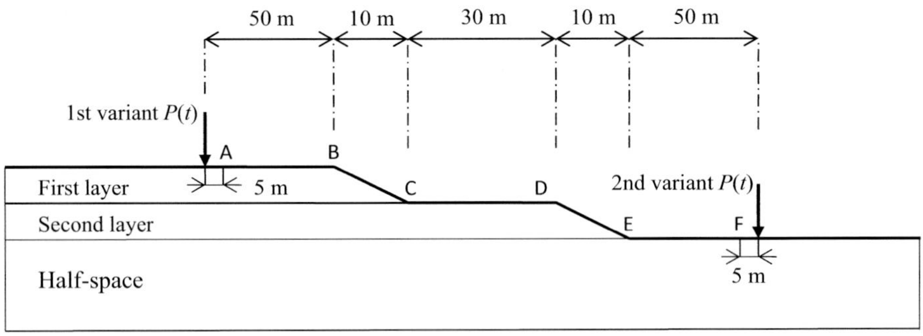

Fig. 9.1 Schematic structure of the model problem

(i) *First layer*:
Elastic modulus $E_1 = 4 \times 10^8$ N/m^2; density $\rho_1 = 1050$ kg/m^3; Poisson's ratio $\nu_1 = 0.29$; longitudinal wave propagation velocity $V_{p1} = 707$ m/s; transverse wave propagation velocity $V_{s1} = 387$ m/s.

(ii) *Second layers*:
Elastic modulus $E_2 = 6 \times 10^8$ N/m^2; density $\rho_2 = 1050$ kg/m^3; Poisson's ratio $\nu_2 = 0.29$; longitudinal wave propagation velocity $V_{p2} = 866$ m/s; transverse wave propagation velocity $V_{s2} = 474$ m/s.

(iii) *Half-space*:
Elastic modulus $E_3 = 8 \times 10^8$ N/m^2; density $\rho_3 = 1050$ kg/m^3; Poisson's ratio $\nu_3 = 0.29$; longitudinal wave propagation velocity $V_{p3} = 1000$ m/s; transverse wave propagation velocity $V_{s3} = 548$ m/s.

Figure 9.2 shows the shape and size of the region in which it is necessary to calculate the dynamic characteristics of wave propagation in a layered structure. The choice of representative volume is carried out according to the criteria and stages presented in Chap. 8.

The lattice for partitioning the representative volume of a three-layer structure into FEs is shown in Fig. 9.3.

The dynamic effect on the system is specified in the form of a nonstationary force, which changes in time according to the law presented in Fig. 8.8 and having a frequency spectrum corresponding to Fig. 8.24. The oscillation source is located at a distance of 50 m to the left or right of the slope (Fig. 9.1).

In a numerical experiment, calculations were carried out for the same system, but when the initial parameters of the layers were varied:

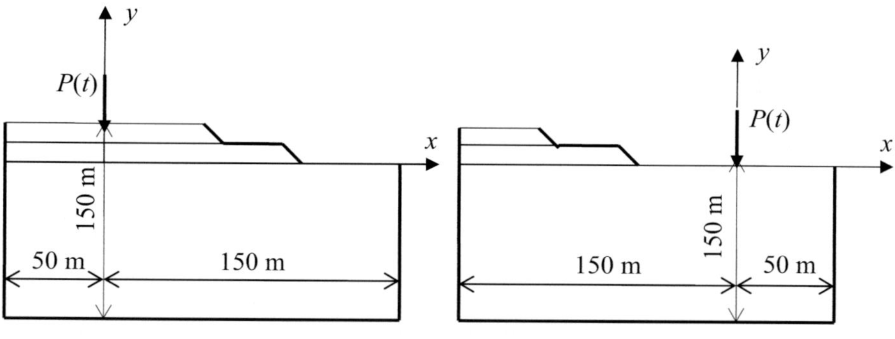

Case 1. Slope is located to the right of the oscillation source

Case 2. Slope is located to the left of the oscillation source

Fig. 9.2 Schematic structure of the shape and size of the area in which the calculation of the dynamic characteristics is necessary

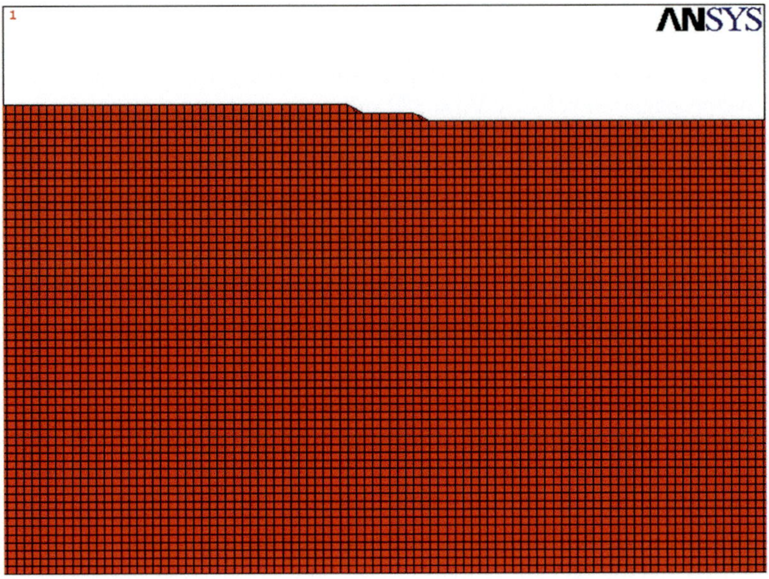

Fig. 9.3 Partitioning lattice into FEs

(i) The ratios of the mechanical characteristics of the material of the layers changed (including the case where all layers have the same mechanical characteristics).
(ii) The geometric characteristics of the layers of the structure and slope were $h_1 = 5$ m; $h_2 = 10$ m or $h_1 = 10$ m; and $h_2 = 5$ m and the angle of inclination varied within 25°–35° (±) 10°.

The most informative for the study of the basic laws of the distribution of waves propagating in the soil are estimates of the dependences of the displacement amplitude on time (the amplitude-time characteristic of oscillations) and distributions of displacements in the vicinity of the slope at different points in time. A description of the patterns of wave distribution is based on the systematization of the results of a numerical experiment for a different ratio of geometric and mechanical characteristics of the structure. The plots and charts show the most characteristic patterns. When waves fall on a slope from different directions, some general patterns and differences are observed, both quantitative and qualitative. A study was made of the influence of the slope on the quantitative and qualitative characteristics of the distribution of oscillations in the vicinity of the slope for various cases of the structure of the slope and the position of the oscillation source in respect to it.

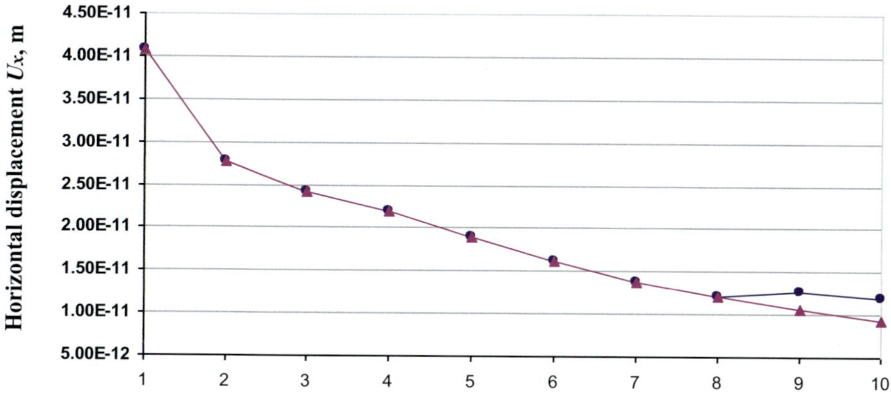

Fig. 9.4 Plots of the maximum amplitudes of displacements U_x at points of section A (5, 0)–B (50, 0); triangles correspond to normal structure without slope; dots correspond to normal structure with slope

9.1.1 Source of Oscillations Disposes to the Left of Slope

The most characteristic distributions of displacement amplitudes in dependence on time (when a wave packet passes through a slope) are presented in Figs. 9.6, 9.7, 9.10, 9.11, 9.12, 9.13, 9.16, and 9.17. In the case of loading the system to the left of the slope (Fig. 3.2, case 1), the maximum displacement near the slope is significantly higher than when the source is located to the right of it.

This is explained by the fact that the intensity of waves, generated in a layered structure without a slope, when the source of vibrations is located on a softer layer, is higher than when it is located on a stiffer layer. Moreover, the half-space of the normal structure is characterized by the finite-resonant nature of the oscillations, and their amplitude substantially depends on the frequency (the ratio of the intrinsic spectrum of the system to the spectrum of the load). An additional factor, determining the increase in the amplitudes of oscillations in the case under consideration, is the presence in the upper half-layer of waves reflected from its inclined edge. Their interaction with the direct field of the source determines the zone of increase in the level of oscillations near the slope.

A quantitative assessment of the influence of the slope was carried out by using comparisons of the maximum displacement amplitudes in various cases. In all cases, there is a stronger influence of the slope on the amplitude of the horizontal displacement than on the vertical. As an example, plots of the maximum displacement amplitudes U_x are present in Fig. 9.4.

Based on a comparison of the results of calculations carried out for different ratios of the mechanical characteristics of the material of the layers, it can be noted that in the case when all layers have the same mechanical characteristics (uniform slope), the influence of the slope on the intensity of the propagating waves is less than in the case of a normal structure. Figure 9.5 presents histograms of maximum amplitudes at point B.

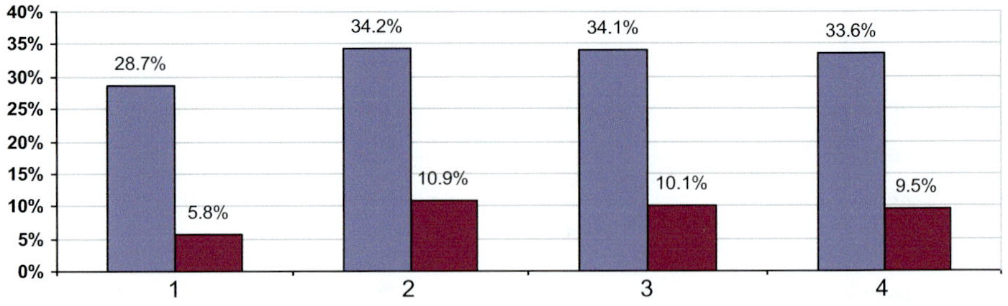

Fig. 9.5 Histograms of the percentage increase in the maximum amplitudes U_x (left columns) and $|\vec{U}|$ (right columns) for point B (50, 0) in various structures: (1) normal structure; (2), (3), (4) homogeneous structures, whose mechanical characteristics of materials are, respectively, equal to the parameters of the first, second, and third layers of the normal structure

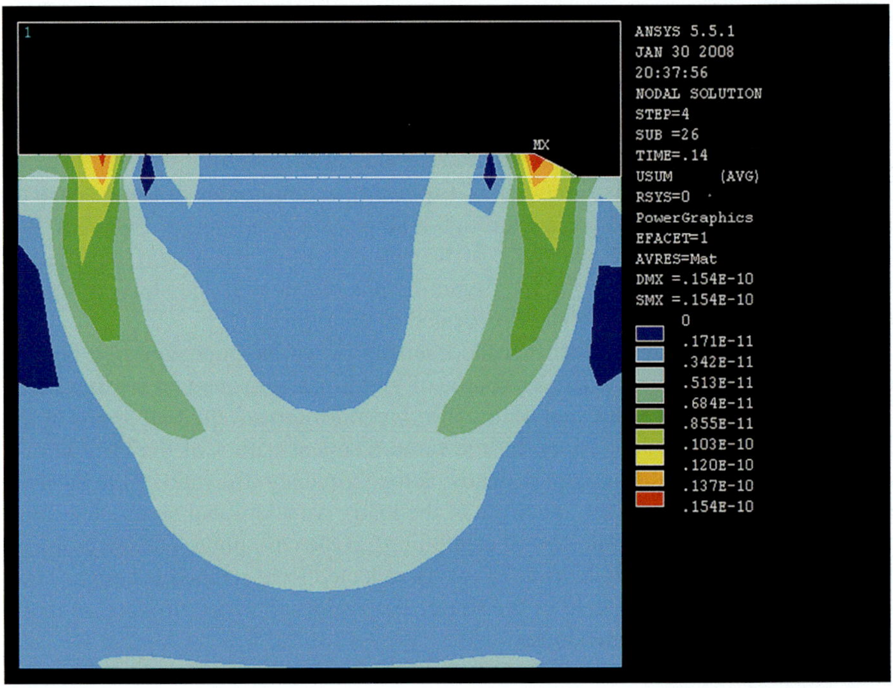

Fig. 9.6 Distribution of displacement amplitudes at the time when wave packet reaches the upper face of the slope ($t = 0.14$ s)

The maximum amplitude of the oscillations is observed in the upper layer of the normal structure above the slope at the time the wave packet reaches the upper face of the slope (Figs. 9.6 and 9.7).

This is explained by the fact that the contributions of the direct wave propagating in the upper layer from the oscillation source and the wave reflected from the edge surface of the half-layer (lateral face of the slope) are summed up namely in this time.

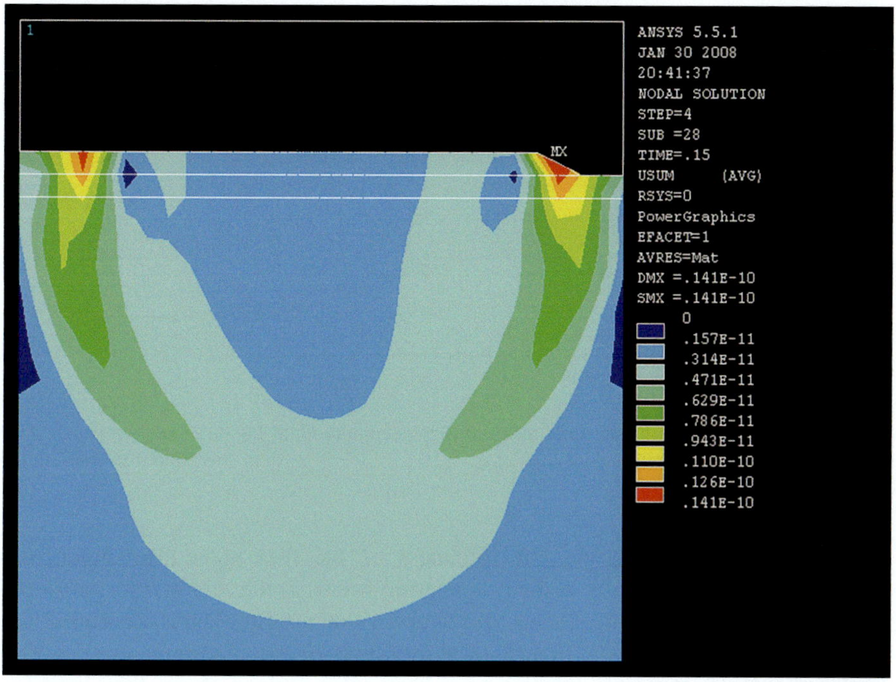

Fig. 9.7 Distribution of displacement amplitudes at the time of the output of the wave packet in the vicinity of the slope ($t = 0.15$ s)

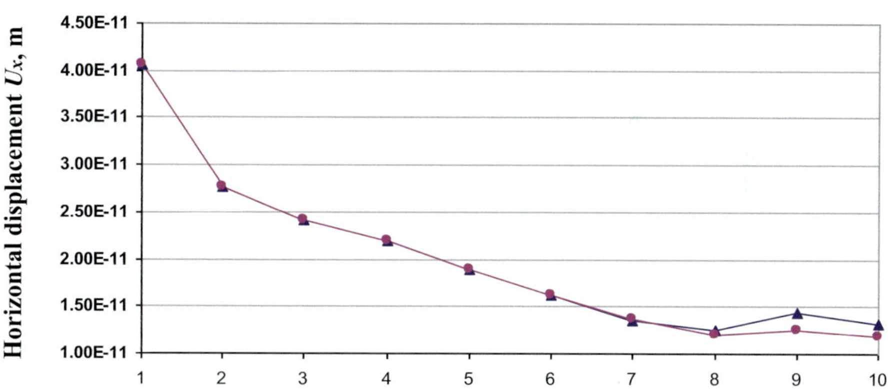

Fig. 9.8 Plots of the maximum amplitudes of displacements U_x at points of section A (5, 0)–B (50, 0); triangles correspond to the inclination angle of 45°; dots correspond to the inclination angle of 26.5°

The amplitude of the oscillations is higher, the greater the angle of inclination of the slope. Figure 9.8 shows plots of the maximum amplitude of displacements U_x at points located on a horizontal surface above the slope for various values of the angle of inclination.

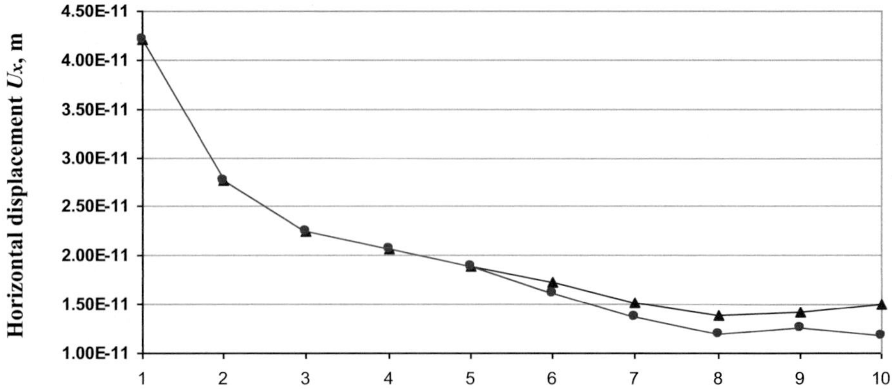

Fig. 9.9 Plots of the maximum amplitudes of displacements U_x at points of section A (5, 0)–B (50, 0); triangles correspond to the case of h_1 = 10 m; h_2 = 5 m; dots correspond to the case of $h_1 = h_2 = 5$ m

Like the inclination angle, the thickness of the first layer also affects the amplitude of the oscillations of the propagating waves in the region near the slope. In the normal structure, the greater the thickness of the first layer, the higher the amplitude of the oscillations. Figure 9.9 shows plots of the maximum amplitude of displacements U_x at points located on a horizontal surface above the slope for various thicknesses of the first layer.

When the leading front of the wave propagating from the source of oscillations reaches the region of exit to the surface of the middle layer of the slope, the oscillation amplitude gradually decreases, as shown in Figs. 9.10 and 9.11.

An increase in the oscillation amplitude is again observed when the propagating waves approach the edge of the middle layer (see Fig. 9.12).

In the vicinity of the edge part of the middle slope, an increase in the intensity of propagating waves is observed (see Fig. 9.13).

In all cases, as previously noted, a stronger influence of the slope on the horizontal component of the displacement vector U_x is observed (see Fig. 9.14).

It is noted that the structure of the model affects the horizontal component of the displacement vector U_x. The plots of the maximum displacement amplitudes U_x at the slope points of the normal structure (Fig. 9.15) show a decrease in the displacement amplitudes of the middle layer in the vicinity of its exit to the surface. If we consider a structure in which the elastic properties of all layers are the same (homogeneous structure), then an increase in the amplitude of horizontal displacement in the same region is observed.

The vertical component of the displacement vector U_y near the surface of the middle layer does not show a decrease in the amplitude of the displacement in the region of exit to the surface of this layer.

In the vicinity of the exit to the surface of the half-space, the oscillation amplitude does not decrease, compared with the middle layer (see Figs. 9.16 and 9.17). This is because the middle layer is in contact with a more rigid half-space.

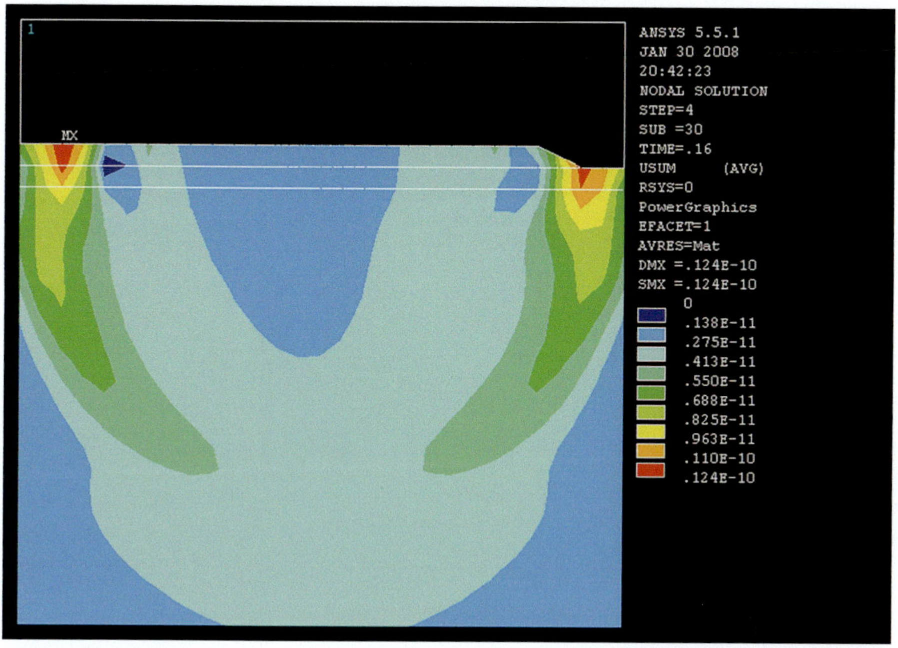

Fig. 9.10 Distribution of displacement amplitudes at the time of the exit of the wave packet from the slope on the surface of the second layer ($t = 0.16$ s)

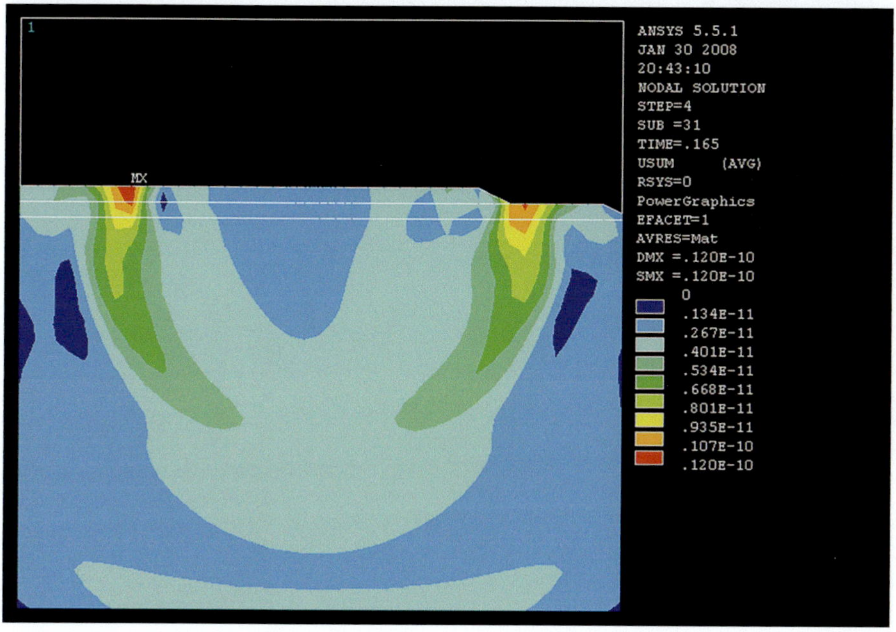

Fig. 9.11 Distribution of displacement amplitudes at the time of propagation of the wave packet on the surface of the second layer ($t = 0.165$ s)

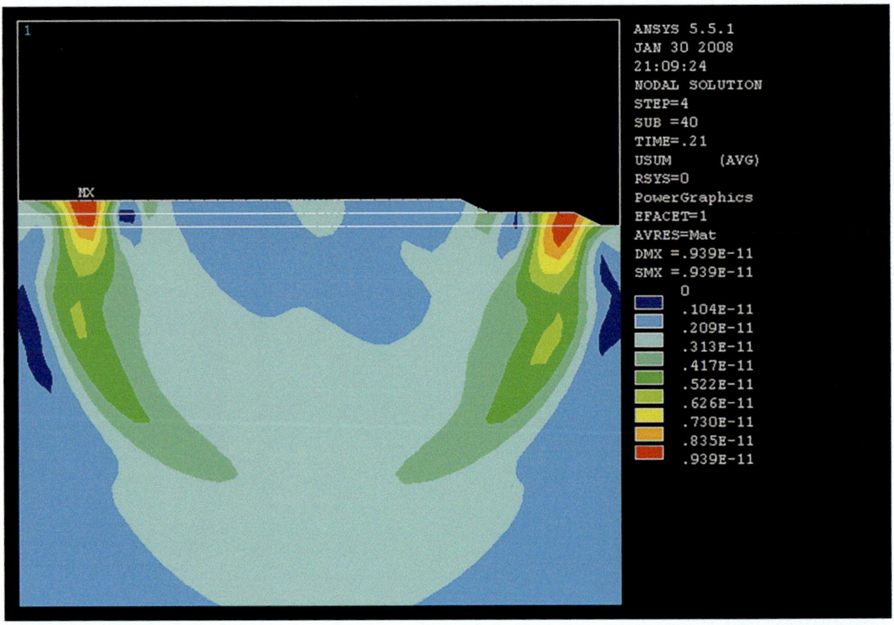

Fig. 9.12 Distribution of displacement amplitudes at the time when the propagating wave packet approaches the edge of the middle layer ($t = 0.21$ s)

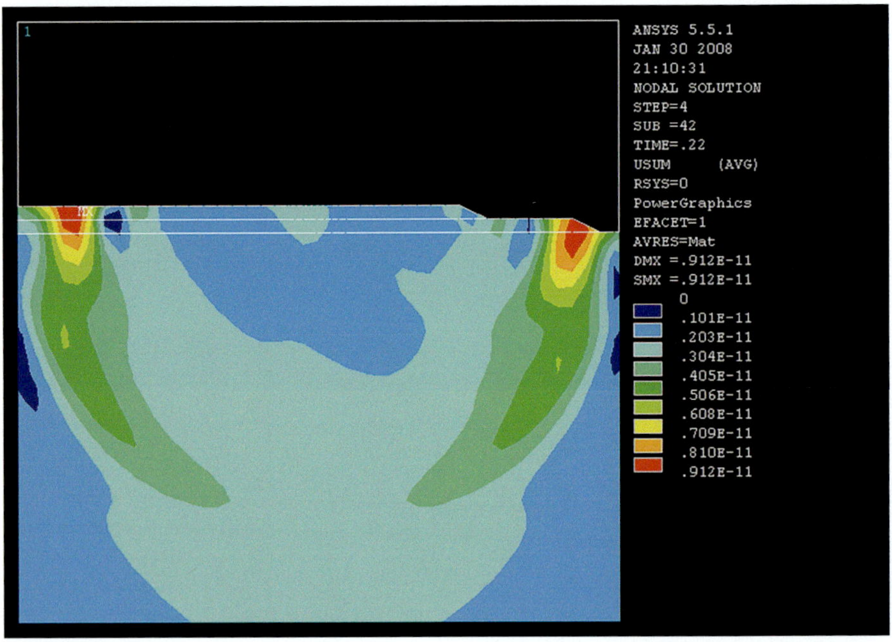

Fig. 9.13 Distribution of the displacement amplitudes at the time of propagation of the wave packet in the vicinity of the slope of the second layer ($t = 0.22$ s)

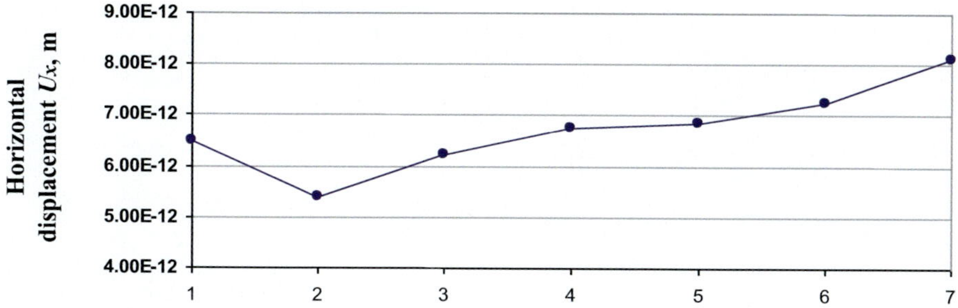

Fig. 9.14 Plot of the maximum amplitudes of the displacements U_x at points on section *C* (60, 0)–*D* (90, 0)

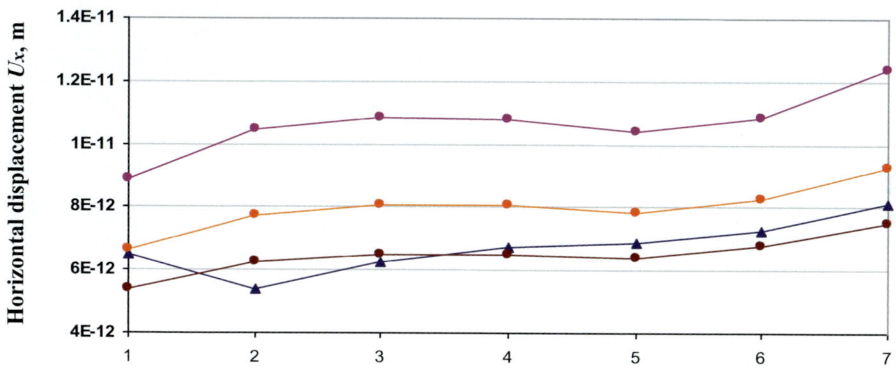

Fig. 9.15 Plots of the maximum displacement amplitudes U_x at points on the stepped section *C* (60, 0)–*D* (90, 0); triangles correspond to the case of normal structure; dots correspond to homogeneous structure with mechanical parameters equal to the parameters of the first, second, and third layers of the normal structure, respectively

At the half-space boundary, an increase in the displacement amplitudes of the horizontal component of the displacement vector U_x in the region of exit from the slope is observed. In the case of a normal structure, the growth of the displacement amplitudes is slower than in the case of a homogeneous structure (see Fig. 9.18).

Changes in the maximum displacement amplitudes on the free surface of a normal structure with a slope depending on the linear coordinate (along the broken line *ABCDEF*) are shown in Fig. 9.19.

9.1.2 Source of Oscillations Disposes to the Right of Slope

The most characteristic distributions of displacement amplitudes at different times (when a wave packet passes through a slope) are presented in Figs. 9.20, 9.24, 9.25, 9.26, 9.27, 9.29, and 9.30. In the case of a wave propagating from a source located to

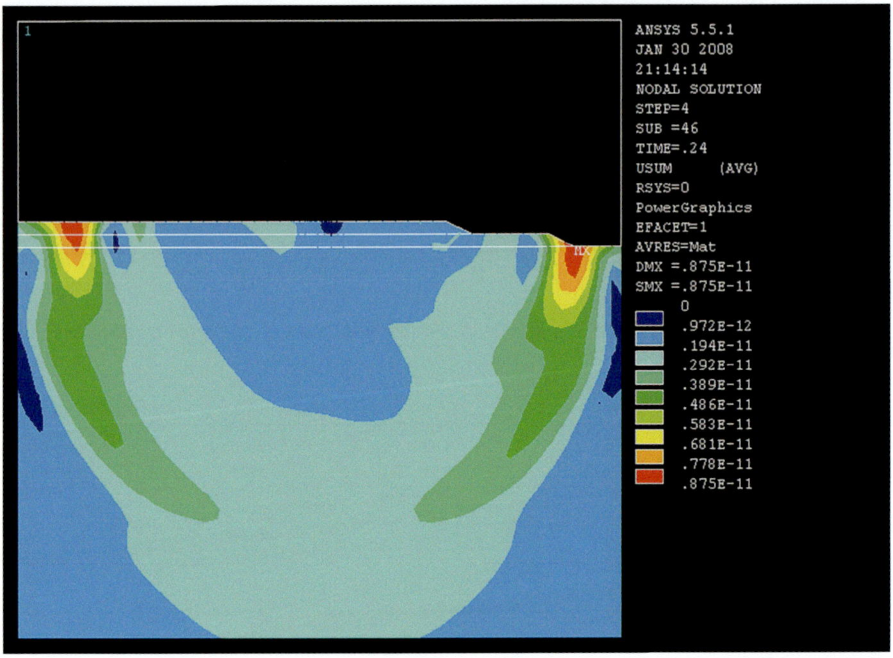

Fig. 9.16 Distribution of the displacement amplitudes at the time of propagation of the wave packet to the surface of the third layer near the slope ($t = 0.24$ s)

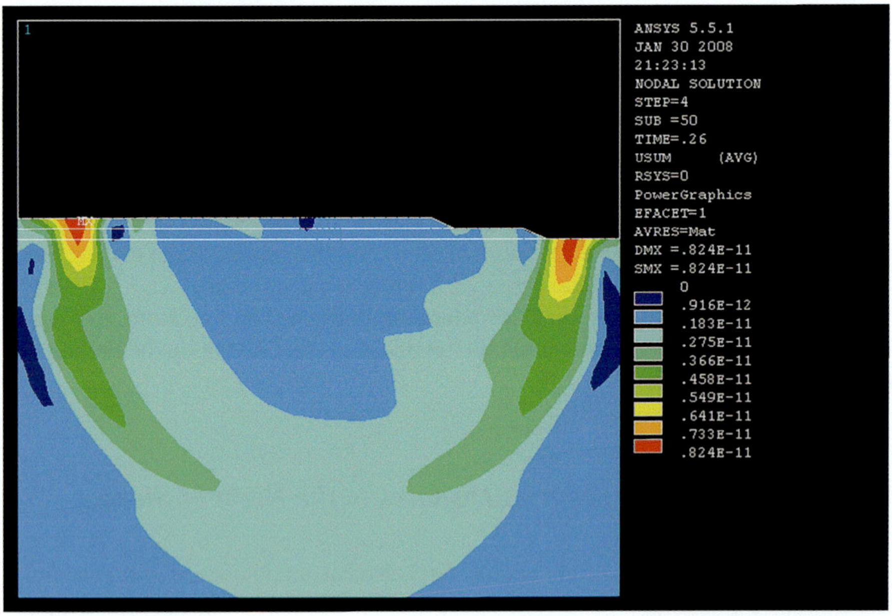

Fig. 9.17 Distribution of the displacement amplitudes at the time of propagation of the wave packet to the surface of the third layer ($t = 0.26$ s)

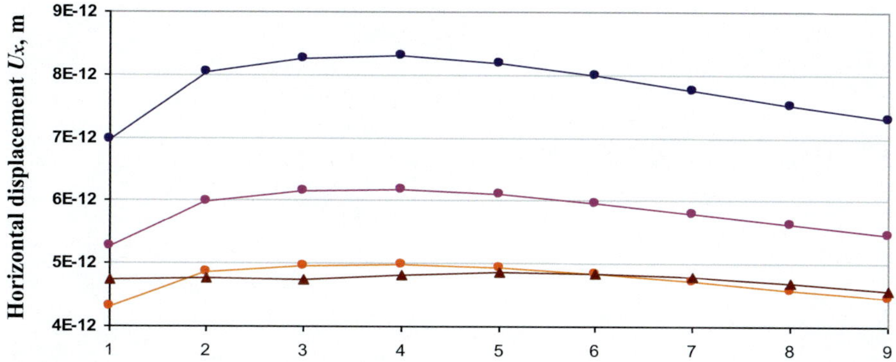

Fig. 9.18 Plots of the maximum displacement amplitudes U_x at points on the surface of the third layer E (100, −5)–F (145, −5); triangles correspond to the case of normal structure; dots correspond to homogeneous structure with mechanical parameters equal to the parameters of the first, second, and third layers of the normal structure, respectively

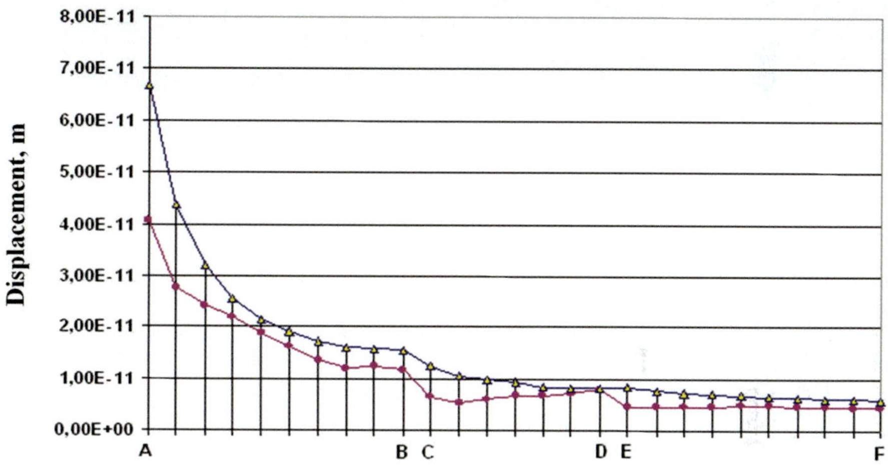

Fig. 9.19 Plots of maximum displacement amplitudes at points on the surface of the first (AB), second (CD), and third layers (EF) of normal structure under loading on the left; triangles correspond to U_y and dots correspond to U_x

the right of the slope (Fig. 9.2, case 2), the oscillation amplitude near the source is lower than when it is located on a softer layer of the structure (on the left—above the slope).

As in the case of the location of the source to the left of the slope, a quantitative assessment of the influence of the slope was carried out based on comparisons of the maximum displacement amplitudes in various cases. In all cases, there is a stronger effect of the slope on the amplitude of horizontal displacement than vertical.

When waves propagate in half-space in the vicinity of the lower surface of the third layer near the slope, a decrease in the amplitude of displacements is observed

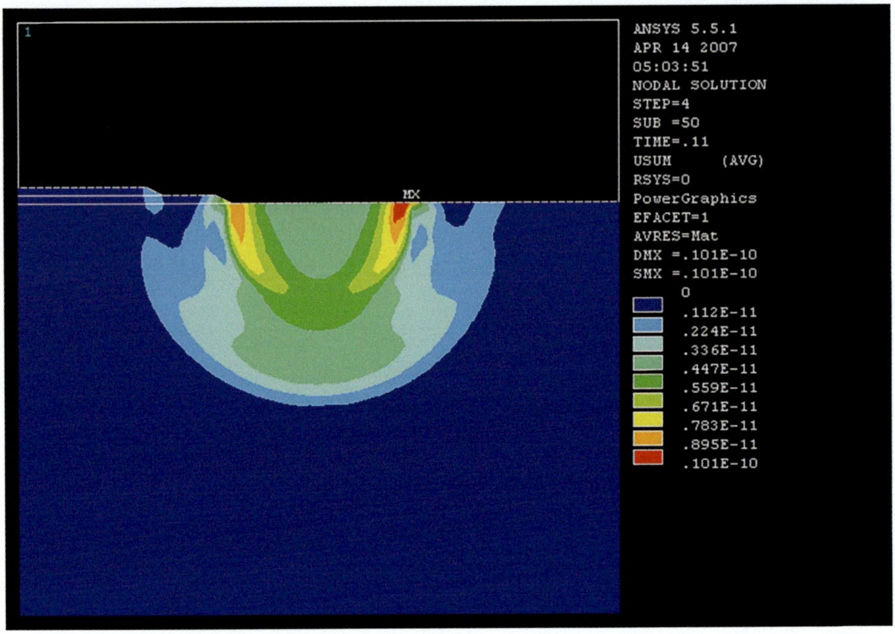

Fig. 9.20 Distribution of the displacement amplitudes at the time of propagation of the wave packet to the surface of the third layer near the slope ($t = 0.11$ s)

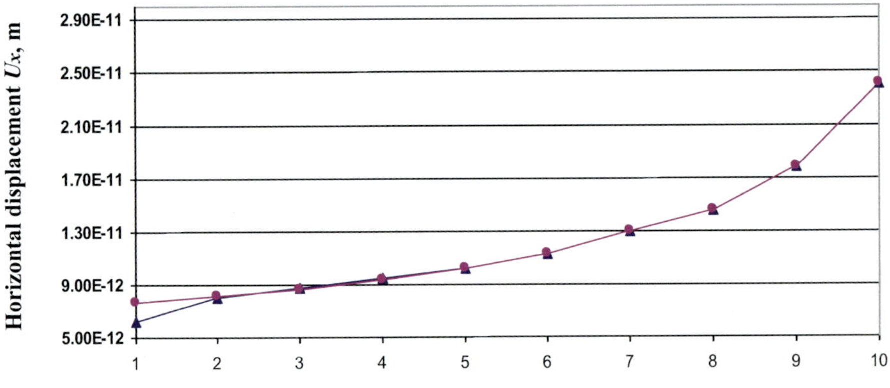

Fig. 9.21 Plots of the maximum amplitudes of displacements U_x at points on the section E (−50, 0)–F (−5, 0); dots correspond to the normal structure without slope; triangles correspond to the normal structure containing the slope

(see Fig. 9.20). Figure 9.21 shows plots of the maximum amplitudes of displacements of the surface of the half-space U_x in the vicinity of the lower part of the slope.

It should be noted that in the case when all layers have the same mechanical characteristics (homogeneous slope), the influence of the slope on the intensity of the propagating waves is greater than in the case of a normal structure. Based on a

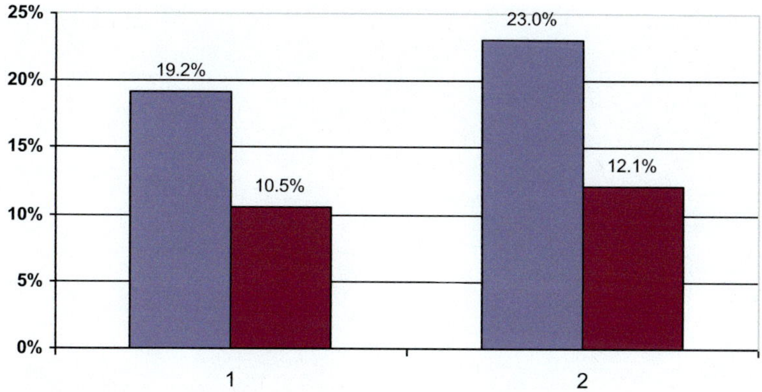

Fig. 9.22 Histograms of the percentage reduction in the maximum amplitudes U_x (left columns) and (right columns) due to the slope for point E in various structures: (1) normal structure; (2) homogeneous structure

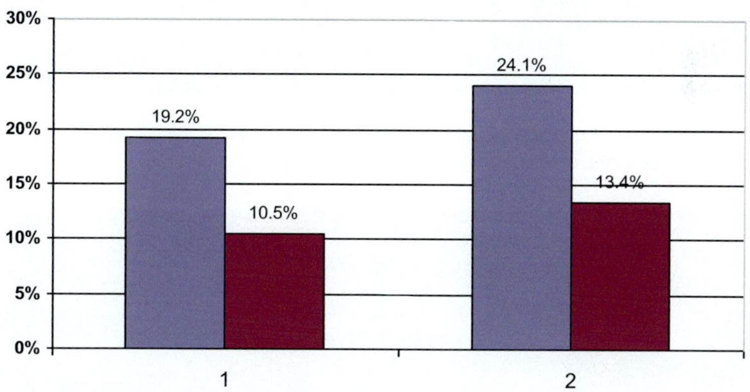

Fig. 9.23 Histograms of the percentage reduction in the maximum amplitudes U_x (left columns) and $|\vec{U}|$ (right columns) due to the slope for point E: (1) inclination angle of the slope is equal to 26.5°; (2) inclination angle of the slope is equal to 45°

comparison of the results of calculations performed for various ratios of the mechanical characteristics of the material of the layers, histograms of the percentage decrease in the maximum amplitudes U_x and U on the influence of the slope can be presented at point E (Fig. 9.22).

As in the case when the oscillation source is located above the slope, a change in the angle of inclination also affects the amplitude of the oscillations of the half-space surface near the lower part of the slope. Figure 9.23 shows histograms of the percentage decrease in the maximum amplitudes U_x and U on the influence of the slope at point E for different inclination angles of the second layer. A change in the thickness of the second layer slightly affects the amplitude of the oscillations of the surface of the half-space near the bottom of the slope.

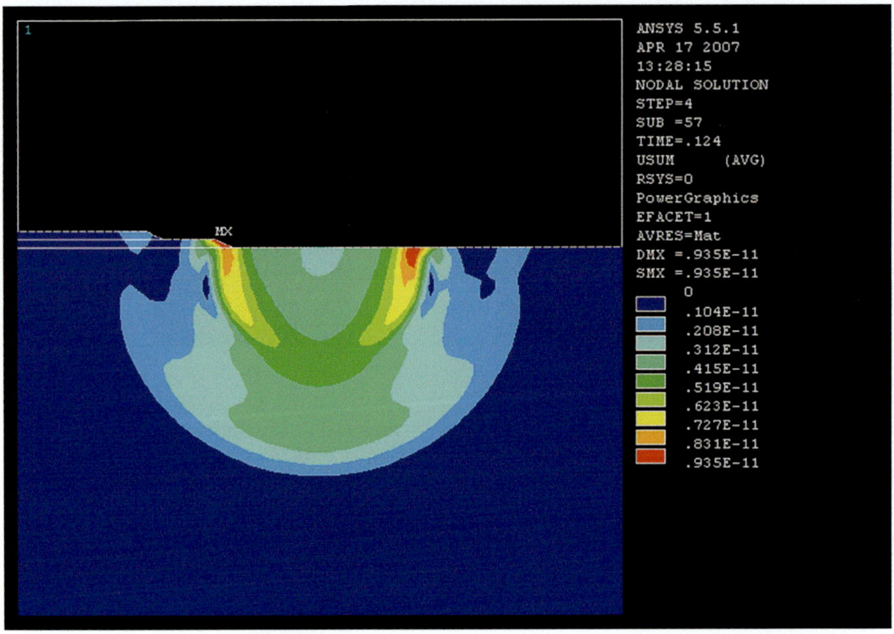

Fig. 9.24 Distribution of the displacement amplitudes at the time of the output of the wave packet in the vicinity of the slope of second layer ($t = 0.124$ s)

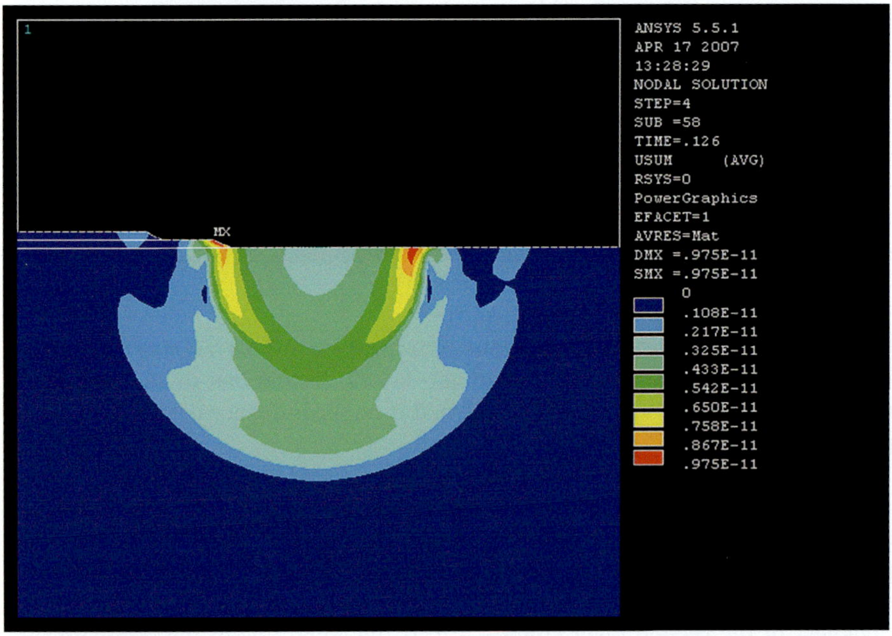

Fig. 9.25 Distribution of displacement amplitudes at the moment the wave packet exits to the upper face of the slope of second layer ($t = 0.126$ s)

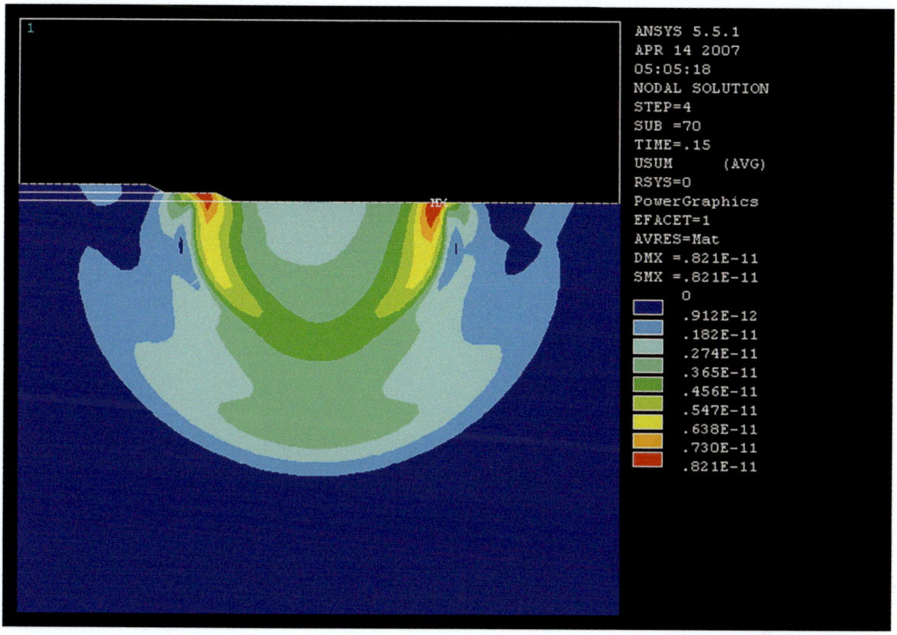

Fig. 9.26 Distribution of the displacement amplitudes at the time of propagation of the wave packet to the surface of the second layer ($t = 0.15$ s)

The maximum oscillation amplitudes are observed near the surface of the slope when a wave packet enters it (see Figs. 9.24 and 9.25).

By reaching the surface of the middle layer, the amplitude of the oscillations significantly decreases (see Figs. 9.26 and 9.27).

Figure 9.28 presents plots of the horizontal components of the displacement vector U_x on the stepped part of the slope for normal and homogeneous structures. It is seen that in the case of the normal structure of the slope layers, the decrease in amplitudes is less than that in the homogeneous structure.

In all layers of the slope, a change in the angle of inclination slightly affects the amplitude of surface vibrations. When the propagating waves reach the upper layer of the slope, a slight increase in the intensity of oscillations in the vicinity of the slope is also observed (see Fig. 9.29).

Upon reaching the surface of the upper layer, the amplitude of the oscillations again decreases (see Fig. 9.30).

Changes in the maximum displacement amplitudes on the surface of various layers in the normal structure along the broken line *ABCDEF* and their dependence on distance from the source of oscillations are shown in Fig. 9.31.

To compare the laws of the distribution of the amplitudes of displacements along the broken surface of the normal structure with the slope, Fig. 9.32 shows graphs of the maximum amplitudes of displacements U on the surface of layers of a normal structure along a broken line *ABCDEF*, when a source of oscillations of the same intensity is located on different sides of the slope at an equal distance from it.

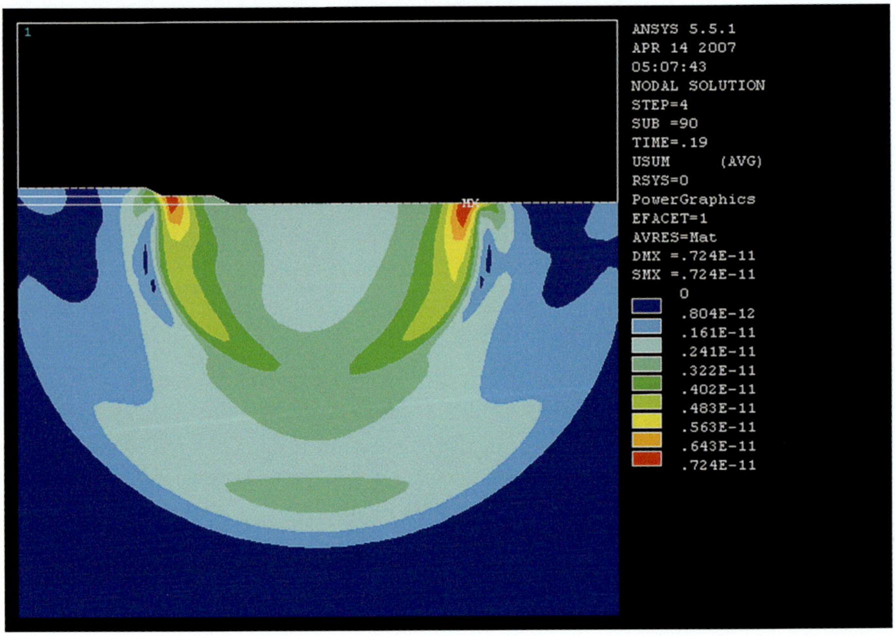

Fig. 9.27 Distribution of the displacement amplitudes at the time of propagation of the wave packet to the surface region near the slope of the first layer ($t = 0.19$ s)

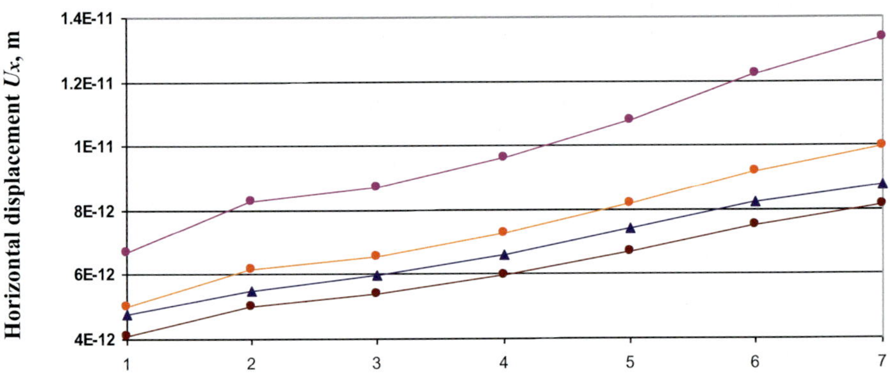

Fig. 9.28 Plots of the maximum displacement amplitudes U_x at points on the stepped section $C(-90, 5)–D(-60, 5)$; triangles correspond to the normal structure and dots correspond to the homogeneous structure with mechanical parameters equal to the parameters of the first, second, and third layers of the normal structure, respectively

Obviously, when the source of the oscillations is below the slope of the normal structure, in its vicinity and above the slope, the amplitude of the oscillations is significantly lower than in the case of the propagation of oscillations from the upper part of the slope. Consequently, if the most probable direction of seismic wave incidence

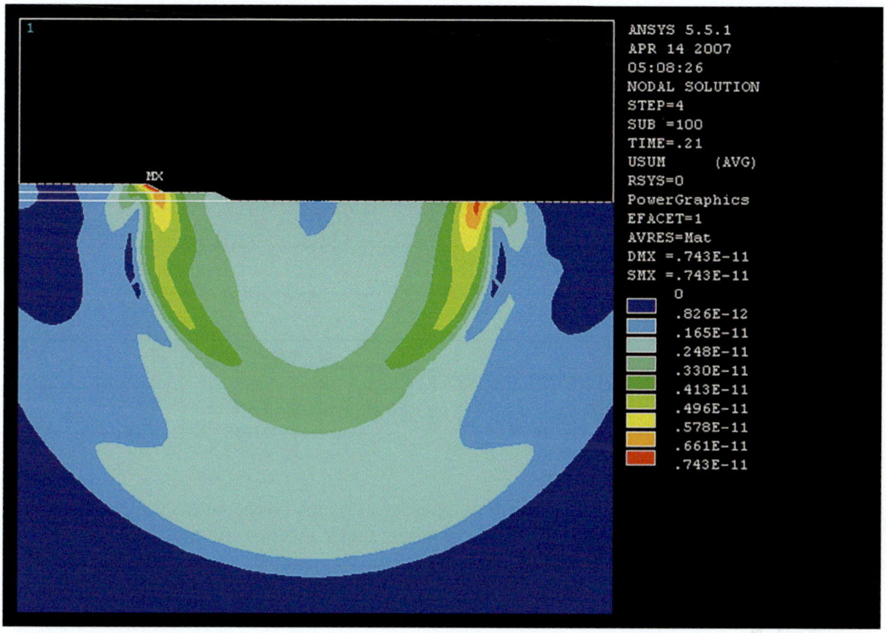

Fig. 9.29 Distribution of the displacement amplitudes at the time of the output of wave packet at the slope region of the first layer ($t = 0.21$ s)

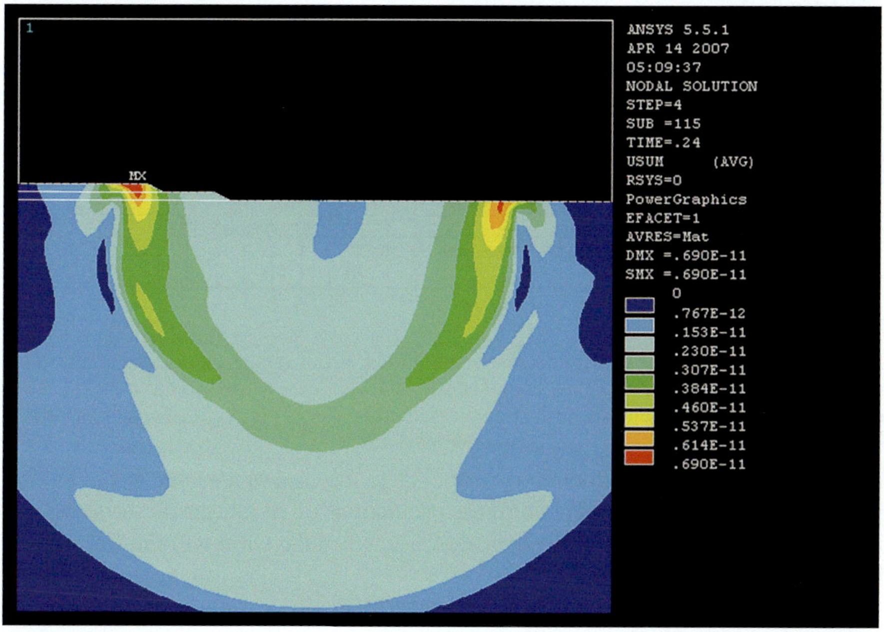

Fig. 9.30 Distribution of the displacement amplitudes at the time of propagation of the wave packet to the surface region of the first layer ($t = 0.24$ s)

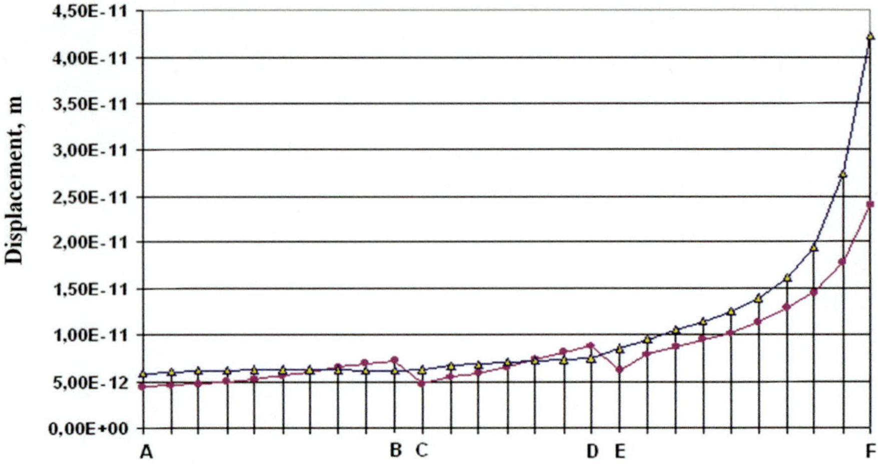

Fig. 9.31 Plots of maximum displacement amplitudes at points on the surface of the first (*AB*), second (*CD*), and third layers (*EF*) of normal structure under loading on the right; triangles correspond to displacements U_y and dots correspond to displacements U_x

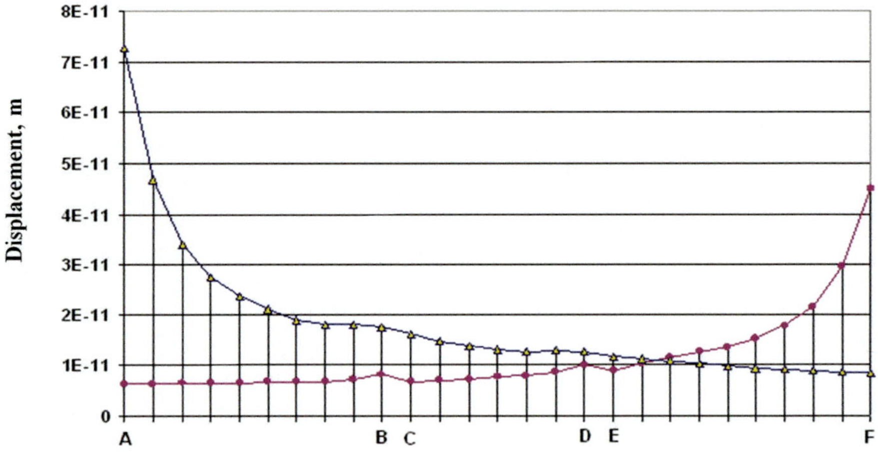

Fig. 9.32 Plots of maximum displacement amplitudes at points on the surface of the first (*AB*), second (*CD*), and third layers (*EF*); triangles correspond to the case when loading on the left hand; dots correspond to the case when loading on the right hand

on a slope of the normal structure corresponds to the case of a source located to the right of the slope, then the effect of these oscillations of all objects on the slope and above it will be lower than in the opposite case, when the wave will fall from top to bottom on the slope.

For a normal structure, an increase in the slope angle significantly affects an increase in the amplitude of oscillations.

9.2 Anomalous Structure (Stiffness of Layers Changes with Depth Not Monotonously)

As noted above, the patterns of the distribution of vibrations in a multilayer half-space of an anomalous structure (without a slope) differ significantly from the case of a normal structure. In this case, two cases of anomalous structure (the first case: "hard–soft–hard" and the second case: soft–hard–soft) also differ significantly from each other. Note that in the first case of the anomalous structure:

(i) The frequency response is monotonically decaying (nonresonant) in nature.
(ii) Propagating surface waves such as Rayleigh waves are virtually absent.
(iii) In the inner, softer (low-speed) layer, a waveguide effect can be observed and the concentration of vibrational energy associated with it, when reaching the slope, can lead to a significant increase in the amplitudes of the impact on the object located nearby.

In the second case of the anomalous structure:

(i) The frequency response is pronounced finite-resonant in nature.
(ii) There are sufficiently energetic surface waves such as Rayleigh waves, which, upon reaching the slope, can determine a significant increase in the level of dynamic impact on an object located near the upper part of the slope.
(iii) An intermediate, more rigid (high-speed) layer can play the role of a screen preventing the propagation of vibrations into half-space and, possibly, significantly reducing the level of influence of propagating vibrations on objects located on it and below it, in the presence of a slope.

Taking into account that the presence of a slope significantly changes the picture of the stress-strain state of structure in the vicinity and behind the slope (in the direction of wave propagation), it is necessary to carry out calculations to identify general patterns of the distribution of displacements in an anomalous layered structure in the vicinity of the slope with different structures (cases 1 and 2) and the different locations of the source of loading in relation to the slope.

In a numerical experiment, as in the case of a normal structure, calculations were performed by varying the initial parameters of the layers, and changing the geometric characteristics of the layers of the structure and slope ($h_1 = 5$ m, $h_2 = 10$ m or $h_1 = 10$ m, $h_2 = 5$ m), and the angle of inclination was varied within 25°–35° (±) 10°.

The study of the basic laws of the distribution of waves propagating in the soil is based on the evaluation of plots of the dependence of the amplitude of displacements on time (the amplitude-time characteristic of oscillations) and distribution of displacements in the vicinity of the slope at different points in time. A description of the patterns of wave distribution is based on the systematization of the results of a numerical experiment for a different ratio of geometric and mechanical characteristics of the structure. The plots and charts show the most characteristic patterns. When waves fall on a slope from different directions, some general patterns and differences are observed, both quantitative and qualitative. A study was made of the influence of the slope on the quantitative and qualitative characteristics of the distribution of vibrations in the vicinity of the slope for various cases of the structure of the slope and the position of the oscillation source relative to it.

9.2.1 Distribution of Layer Stiffness in Depth: "Hard–Soft–Hard" with Oscillation Source Disposed to the Left or Right of Slope

For a three-layer structure of the anomalous structure "hard–soft–hard," the mechanical properties of the material are determined by the following basic parameters:

(i) *First layer*:
Elastic modulus $E_1 = 8 \times 10^8$ N/m^2; density $\rho_1 = 1050$ kg/m^3; Poisson's ratio $\nu_1 = 0.29$; longitudinal wave propagation velocity $V_{p1} = 1000$ m/s; transverse wave propagation velocity $V_{s1} = 548$ m/s.

(ii) *Second layers*:
Elastic modulus $E_2 = 4 \times 10^8$ N/m^2; density $\rho_2 = 1050$ kg/m^3; Poisson's ratio $\nu_2 = 0.29$; longitudinal wave propagation velocity $V_{p2} = 707$ m/s; transverse wave propagation velocity $V_{s2} = 387$ m/s.

(iii) *Half-space*:
Elastic modulus $E_3 = 8 \times 10^8$ N/m^2; density $\rho_3 = 1050$ kg/m^3; Poisson's ratio $\nu_3 = 0.29$; longitudinal wave propagation velocity $V_{p3} = 1000$ m/s; transverse wave propagation velocity $V_{s3} = 548$ m/s.

Case 1: Load to the Left of the Slope

The most characteristic diagrams of the distribution of displacement amplitudes at different times (when a wave packet passes through a slope) are presented in Figs. 9.36, 9.37, 9.40, 9.41, 9.42, and 9.43.

In the case of loading the system to the left of the slope (Fig. 9.2, case 1), a rapid decrease in the vibration amplitudes in the surface layer is observed by moving away from the vibration source, at its sufficiently high level in the intermediate, softer layer.

A quantitative assessment of the influence of the slope was carried out on the base of comparisons of the maximum displacement amplitudes in various cases. As an example, Fig. 9.33 shows plots of the maximum displacement amplitudes U_x on the surface of the upper, hard layer from the oscillation source to its edge.

It can be seen that for the anomalous structure "hard–soft–hard," the influence of the slope is manifested only in close proximity to it (at a distance of the order of the thickness of the surface layer), at the time when the wave packet reaches the upper face of the slope (Fig. 9.34).

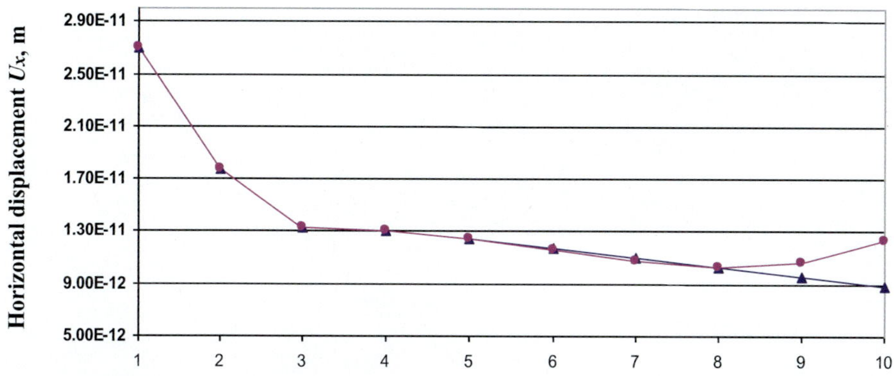

Fig. 9.33 Plots of the maximum amplitudes of displacements U_x at points on section *A* (5, 0)–*B* (50, 0); triangles correspond to the anomalous structure "hard–soft–hard" without a slope and dots correspond to the anomalous structure "hard–soft–hard" containing the slope

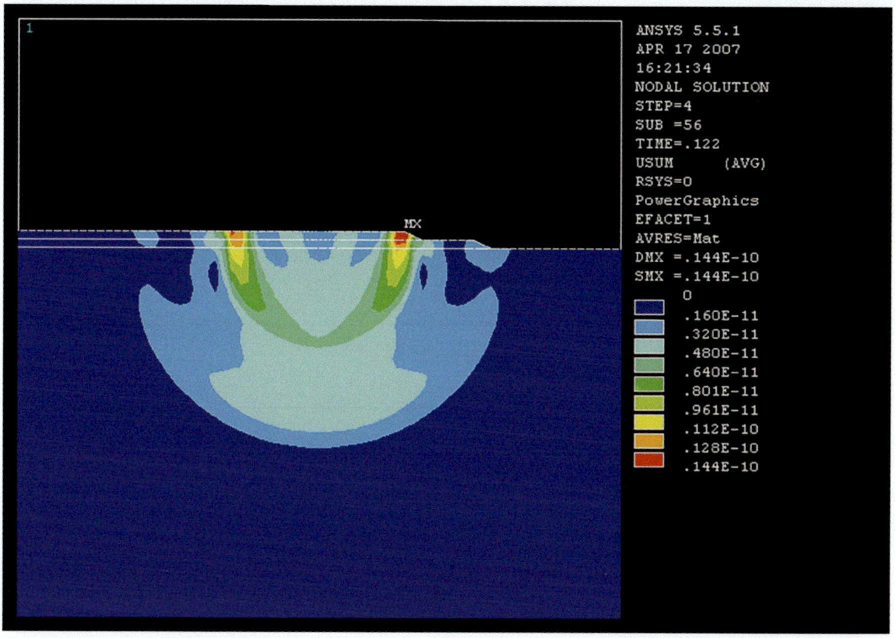

Fig. 9.34 Distribution of displacement amplitudes at the time when the wave packet reaches the upper face of the slope ($t = 0.122$ s)

When the waves exit into the middle, soft layer, there is a significant increase in the amplitude of oscillations in the slope region at the time when the wave packet reaches the lower soft layer (Fig. 9.35).

As in the case of a normal structure, an increase in the inclination angle of the slope determines an increase in the oscillation amplitudes near the edge surface of

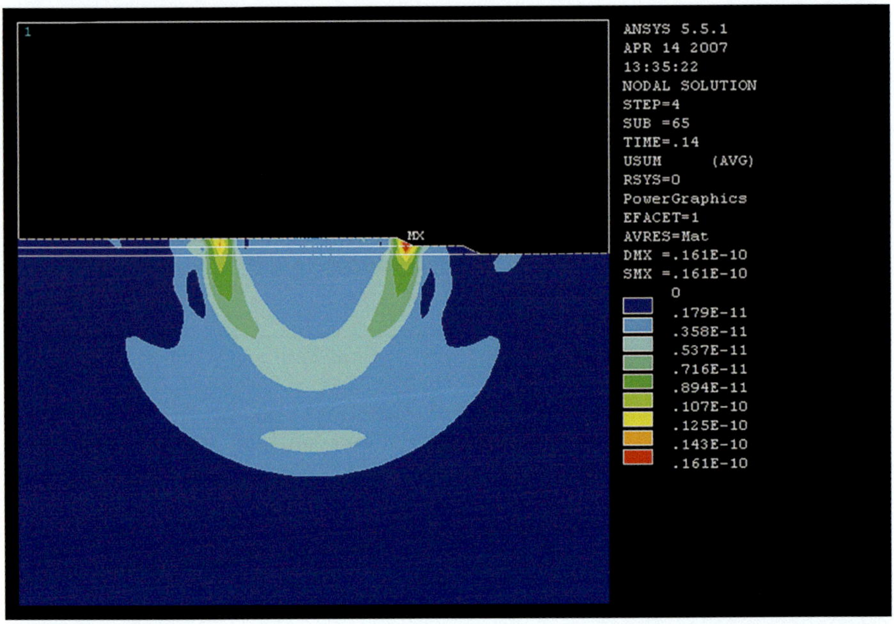

Fig. 9.35 Distribution of the displacement amplitudes at the time when the wave packet reaches the lower soft layer ($t = 0.14$ s)

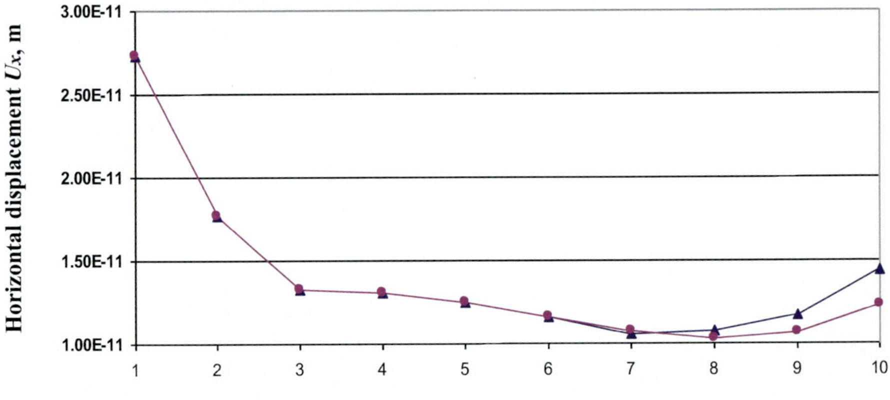

Fig. 9.36 Plots of the maximum values of displacements U_x at points on the section A (5, 0)–B (50, 0) of the anomalous structure "hard–soft–hard"; triangles correspond to an inclination angle of 45° and dots correspond to an inclination angle of 26.5°

the corresponding layer. Figure 9.36 shows plots of the maximum displacement amplitudes U_x at points located on a horizontal surface above the slope for various values of the angle of inclination.

In the anomalous structure, a significant effect of the thickness of the upper layer on the amplitude of oscillations of the propagating waves is observed. The greater

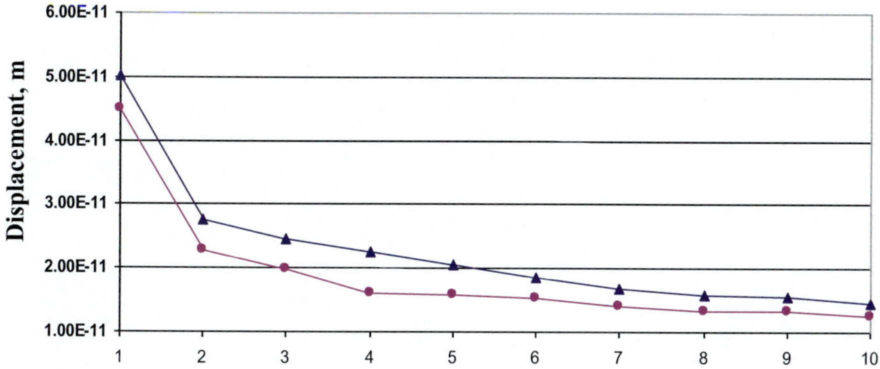

Fig. 9.37 Plots of the maximum values of the displacements $|\vec{U}|$ at the points in section *A* (5, 0)–*B* (50, 0) of the anomalous structure "hard–soft–hard"; triangles correspond to the case of $h_1 = h_2 = 5$ m and dots correspond to the case of $h_1 = 10$ m and $h_2 = 5$ m

its thickness (or higher stiffness), the lower the amplitude of the vibrations in it, and in the whole structure. Figure 9.37 shows graphs of the maximum amplitude of displacements $|\vec{U}|$ at points located on a horizontal surface above the slope for various thicknesses of the upper layer.

After the waves reach the surface of the middle layer of the slope, the amplitude of the oscillations slowly decreases (see Figs. 9.38 and 9.39).

When propagating waves approach the edge of the middle layer, an increase in the amplitude of oscillations is again observed (see Fig. 9.40).

When the wave packet enters the half-space, a noticeable decrease in the oscillation amplitude is observed (see Fig. 9.41).

Changes in the maximum displacement amplitudes on the surface of various slope layers along the broken line *ABCDEF* in an anomalous structure with a distribution of layer stiffnesses from top to bottom: "hard–soft–hard" are presented in the plots of Fig. 9.42.

Case 2: Load to the Right of the Slope

The most characteristic distributions of displacement amplitudes at different times (when a wave packet passes through a slope) are presented in Figs. 9.43, 9.45, and 9.46.

As noted above, in the case of wave propagation from a load source located to the right of the slope (Fig. 9.2, case 2), the oscillation amplitude near the load is lower than when wave propagates from a source located to the left of the slope.

In contrast to the case in which the oscillation source is located to the left of the slope, an equal influence of the slope on the change in the amplitudes of both horizontal and vertical displacements is observed. When the wave packet in the half-space reaches the region of the lower surface of the slope, a slight decrease in the amplitude of the displacements is observed (see Fig. 9.43).

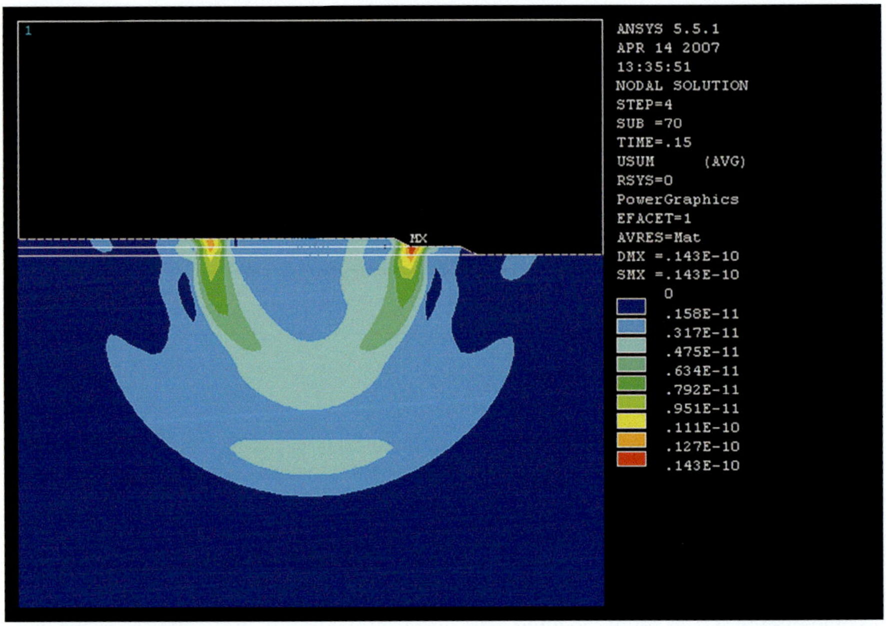

Fig. 9.38 Distribution of the displacement amplitudes at the time when the wave packet reaches the surface of the second layer ($t = 0.15$ s)

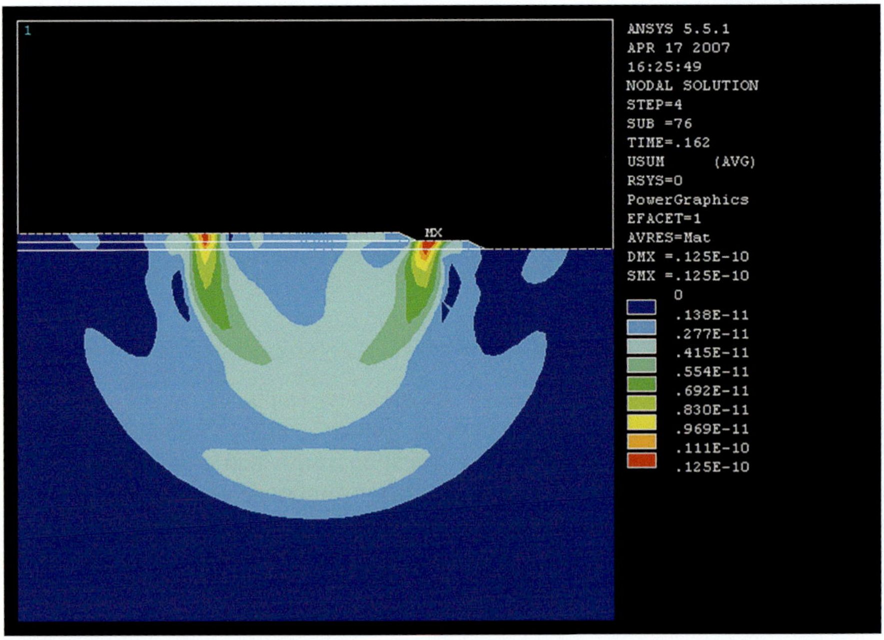

Fig. 9.39 Distribution of the displacement amplitudes at the time of propagation of the wave packet on the surface of the second layer ($t = 0.162$ s)

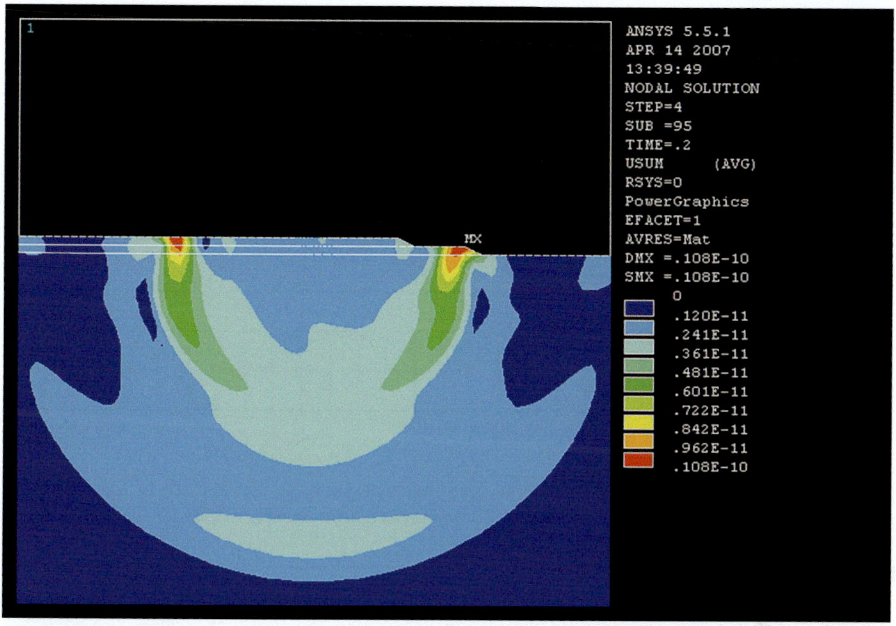

Fig. 9.40 Distribution of displacement amplitudes at the moment of approaching the wave packet to the edge of the middle layer ($t = 0.2$ s)

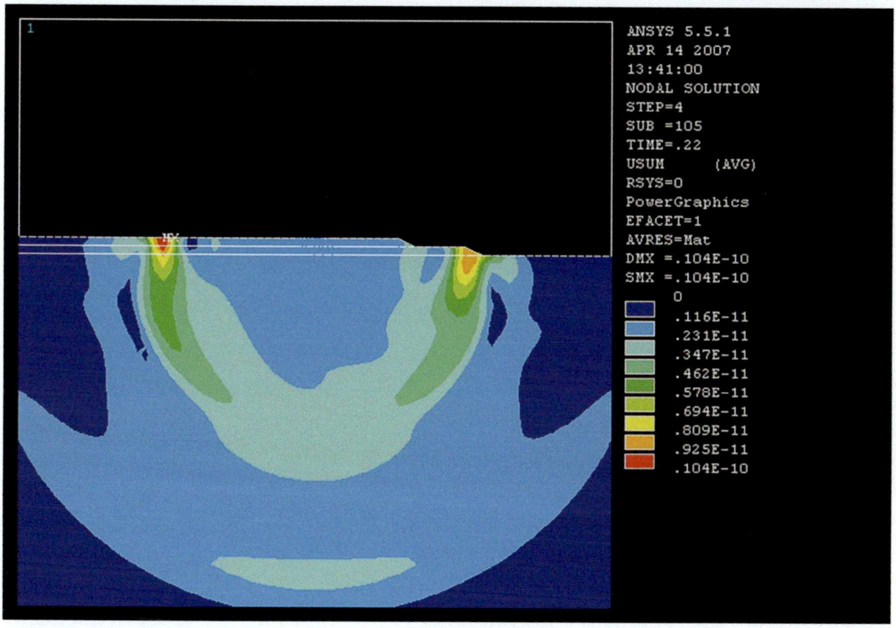

Fig. 9.41 Distribution of the displacement amplitudes at the time when the wave packet enters the half-space ($t = 0.22$ s)

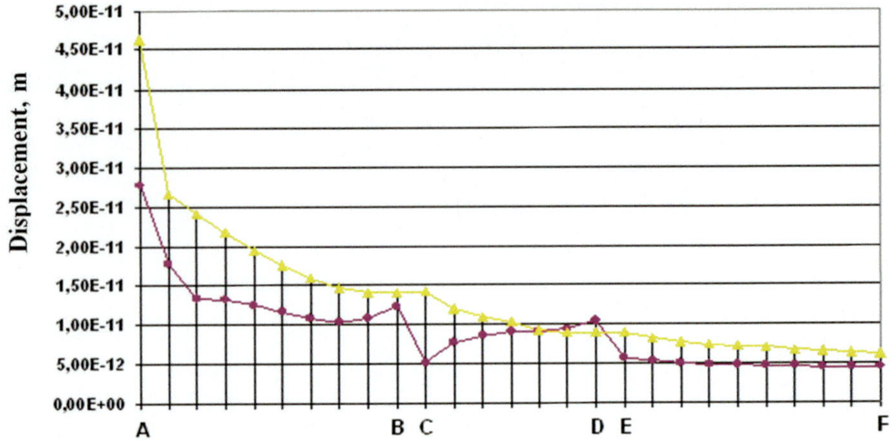

Fig. 9.42 Plots of maximum displacement amplitudes at points on the surface of the first (*AB*), second (*CD*), and third layers (*EF*) of the anomalous structure "hard–soft–hard"; triangles correspond to the U_y and dots correspond to the U_x

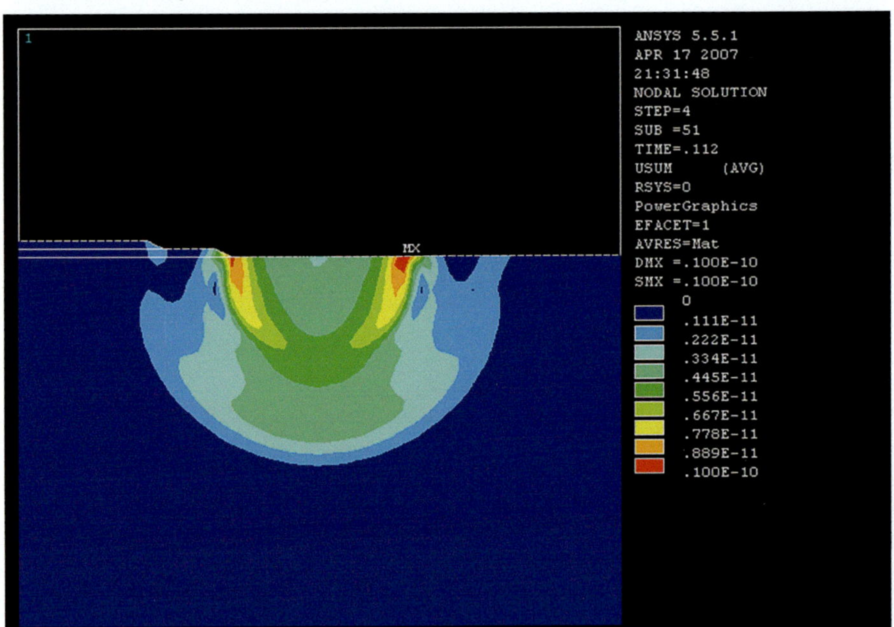

Fig. 9.43 Distribution of the displacement amplitudes at the time when the wave packet attains the surface of the third layer near the slope ($t = 0.112$ s)

A quantitative assessment of the influence of the slope on the change in the maximum amplitudes of displacements of the surface of the half-space U_x in the vicinity of the lower part of the slope is shown in Fig. 9.44.

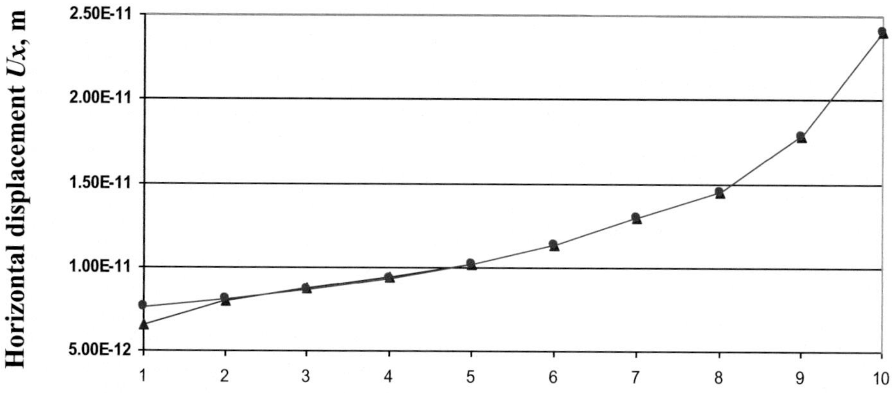

Fig. 9.44 Plots of the maximum amplitudes of displacements U_x at points on the section E (−50, 0)–F (−5.0); triangles correspond to the anomalous structure "hard–soft–hard" without a slope and dots correspond to the anomalous structure "hard–soft–hard" containing a slope

When waves propagate in the region of the second layer of slope, an increase in their intensity is observed. The maximum oscillation amplitudes are observed in the region of the edge surface of the slope (see Fig. 9.45).

When the propagating waves reach the upper layer of the slope, a slight increase in the intensity of oscillations in the vicinity of the slope is also observed (see Fig. 9.46).

The changes in the maximum amplitudes of displacements on the surface of various layers along the broken line *ABCDEF* in the anomalous structure and their dependence on the distance from the vibration source are shown in the plots of Fig. 9.47.

Figure 9.48 shows plots of the maximum displacement amplitudes on the surface of layers of a normal structure with a vibration source of the same intensity located on different sides of the slope at an equal distance from it.

As can be seen from the plots, the amplitude of the oscillations near the load on the half-space is slightly lower than when the load is located on the upper layer. Equal amplitudes are observed with increasing thickness of the upper layer. The vibration levels near the surface of the edge of the inner layer of the structure are practically equal for both cases of the location of the vibration source. Above the slope, the levels of the influence of propagating oscillations are significantly higher when the load is located above the slope (left). That is, for an anomalous structure, the level of influence on the objects located on or above the slope is greater when the load is located to the left of the slope (above).

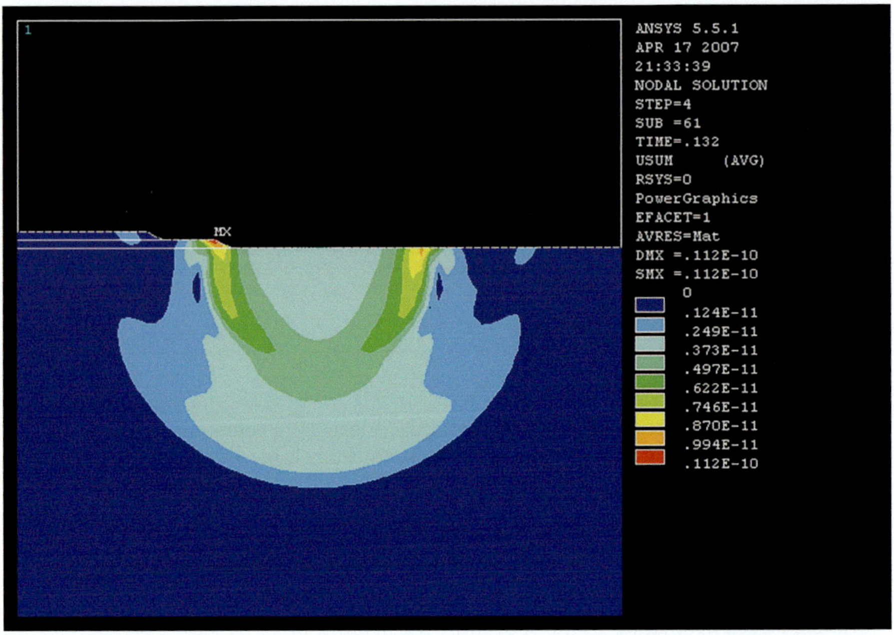

Fig. 9.45 Distribution of the displacement amplitudes at the time, when the wave packet attains the upper face of the second layer of slope ($t = 0.32$ s)

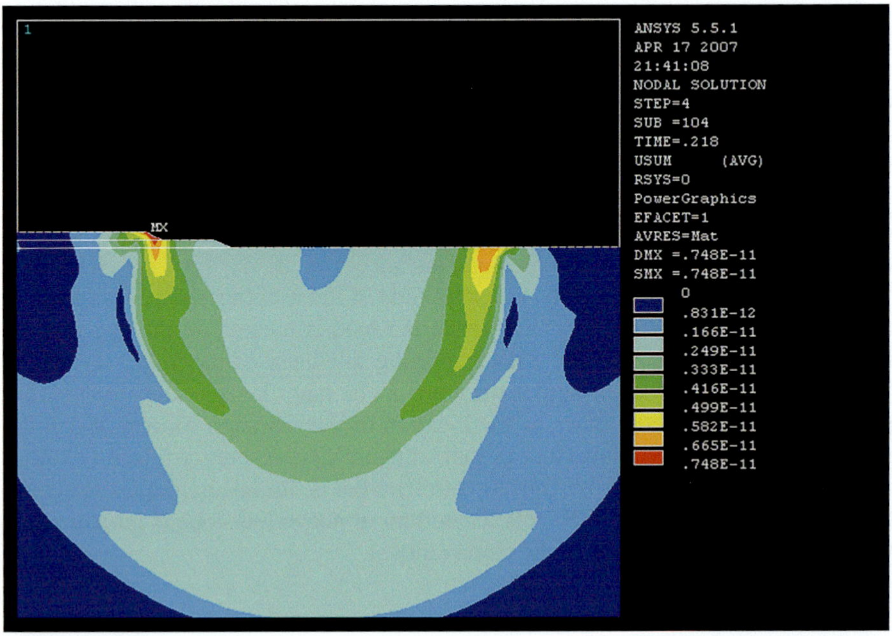

Fig. 9.46 Distribution of the displacement amplitudes at the time when the wave packet attains the vicinity of the first layer of slope ($t = 0.218$ s)

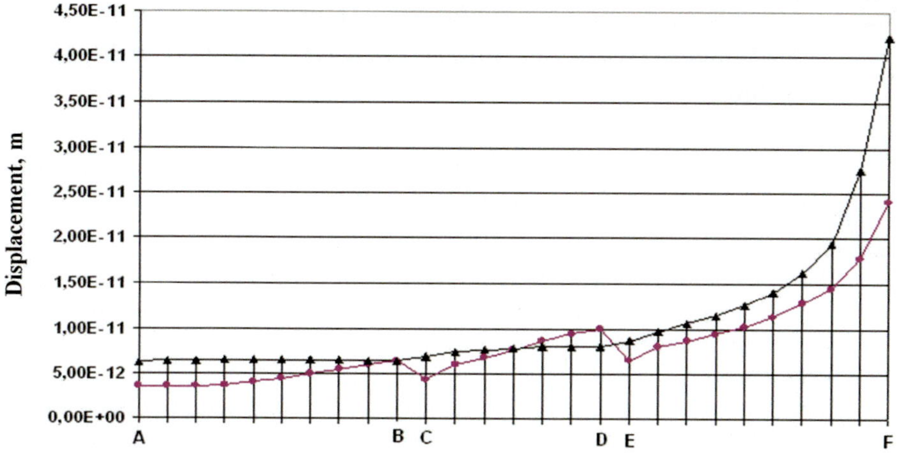

Fig. 9.47 Plots of maximum displacement amplitudes at points on the surface of the first (*AB*), second (*CD*), and third layers (*EF*) of the anomalous structure "hard–soft–hard"; triangles correspond to displacements U_y and dots correspond to displacements U_x

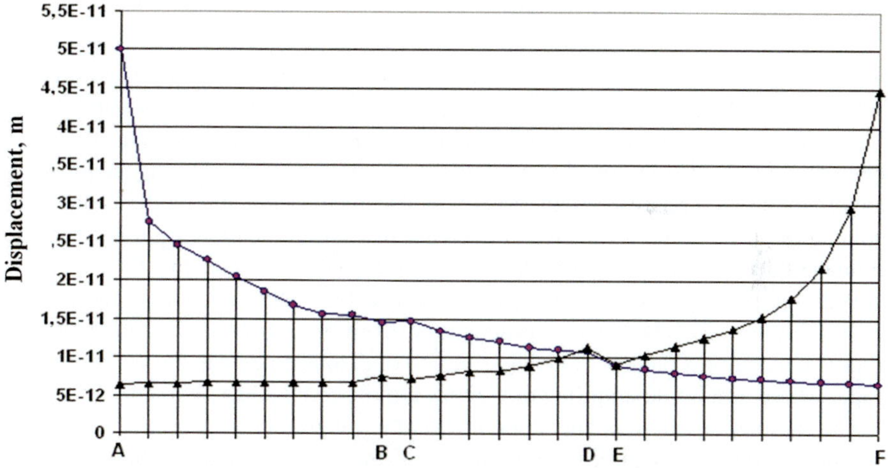

Fig. 9.48 Plots of maximum displacement amplitudes at points on the surface of the first (*AB*), second (*CD*), and third layers (*EF*) of the anomalous structure "hard–soft–hard"; triangles correspond to the case of loading on the right and dots correspond to the case of loading on the left

9.2.2 *Distribution of Layer Stiffness in Depth: "Soft–Hard–Soft" with Oscillation Source Located to the Left or Right of Slope*

For a three-layer structure of the anomalous structure "soft–hard–soft," the mechanical properties of the material are determined by the following basic parameters:

(i) *First layer*:
Elastic modulus $E_1 = 4 \times 10^8$ N/m^2; density $\rho_1 = 1050$ kg/m^3; Poisson's ratio $\nu_1 = 0.29$; longitudinal wave propagation velocity $V_{p1} = 707$ m/s; transverse wave propagation velocity $V_{s1} = 387$ m/s.

(ii) *Second layers*:
Elastic modulus $E_2 = 8 \times 10^8$ N/m^2; density $\rho_2 = 1050$ kg/m^3; Poisson's ratio $\nu_2 = 0.29$; longitudinal wave propagation velocity $V_{p2} = 1000$ m/s; transverse wave propagation velocity $V_{s2} = 548$ m/s.

(iii) *Half-space*:
Elastic modulus $E_3 = 4 \times 10^8$ N/m^2; density $\rho_3 = 1050$ kg/m^3; Poisson's ratio $\nu_3 = 0.29$; longitudinal wave propagation velocity $V_{p3} = 707$ m/s; transverse wave propagation velocity $V_{s3} = 387$ m/s.

Case 1: Load to the Left of the Slope

The most characteristic distributions of displacement amplitudes at different times (when a wave packet passes through a slope) are presented in Figs. 9.49 and 9.51.

For this structure, there is a significant difference in the quantitative and qualitative regularities of the distribution of wave fields in a layered structure. During the propagation of vibrations, a relatively high level of displacement vector amplitudes is observed in the upper layer of the structure; its decrease in the middle, more rigid layer; and a significant decrease in amplitudes in the underlying half-space (see Fig. 9.49).

With the propagation of vibrations in the structure, there is a slight decrease in the amplitudes of surface waves propagating from the source to the left (from the slope), at a complex pattern of the distribution of displacements in the vicinity of the slope, where amplitude amplification zones are observed (Figs. 9.50 and 9.51).

The inclination angle of the slope affects the magnitude of the reflected waves and the degree of growth of the amplitudes of the displacements in the vicinity of the slope upon reaching its wave packet. Figure 9.52 presents plots of the maximum amplitudes of displacements U_x at points located on a horizontal surface above the slope for various values of the angle of inclination.

In the structure of the anomalous "soft–hard–soft" structure, a significant effect of layer thicknesses on the vibration amplitudes in them is also observed. In particular, the greater the thickness of the upper, soft layer, the higher the amplitude of the oscillations in it. Figure 9.53 presents plots of the maximum amplitude of displacements U at points located on a horizontal surface above the slope at various thicknesses.

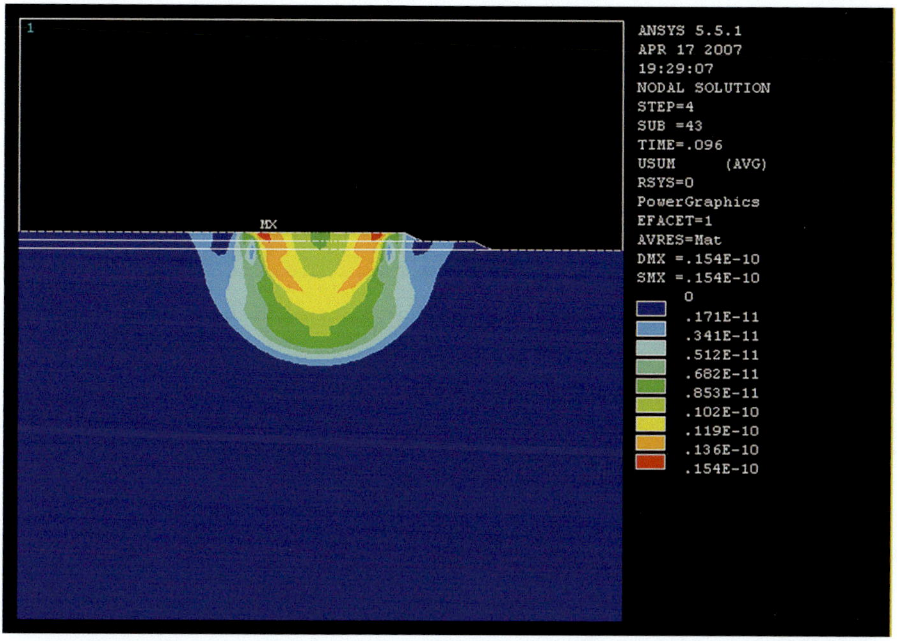

Fig. 9.49 Distribution of the displacement amplitudes at the time of propagation of the wave packet from the source along the surface of first layer ($t = 0.096$ s)

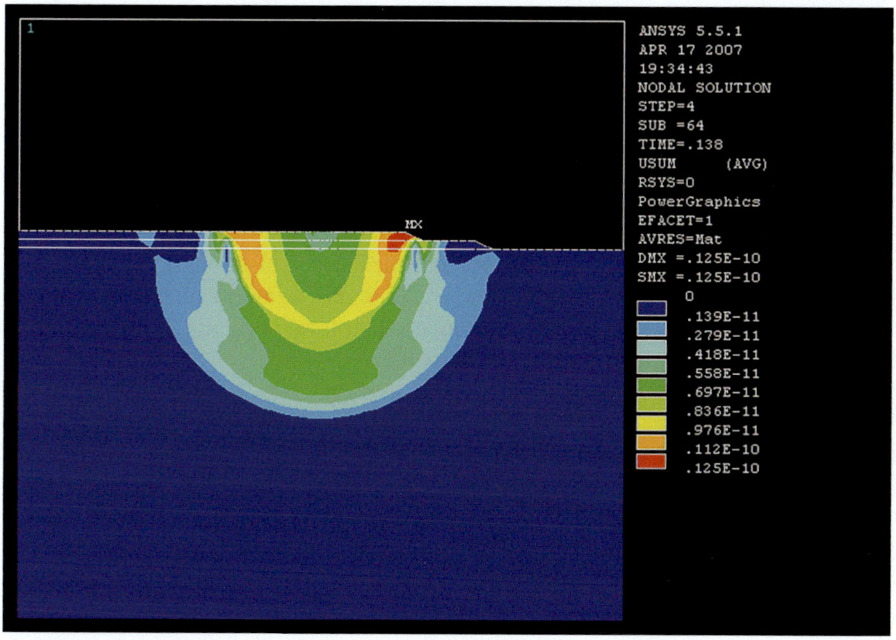

Fig. 9.50 Distribution of the displacement amplitudes at the time when the wave packet reaches the upper face of the slope ($t = 0.138$ s)

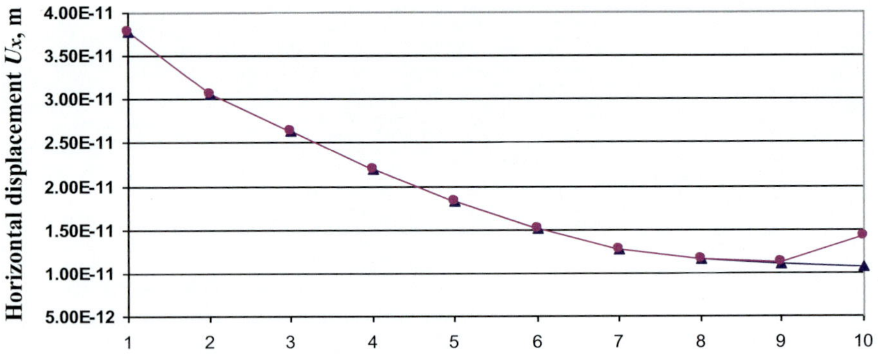

Fig. 9.51 Plots of the maximum amplitudes of displacements U_x at points of section *A* (5, 0)–*B* (50, 0); triangles correspond to the anomalous structure "soft–hard–soft" without a slope and dots correspond to the anomalous structure "soft–hard–soft" containing a slope

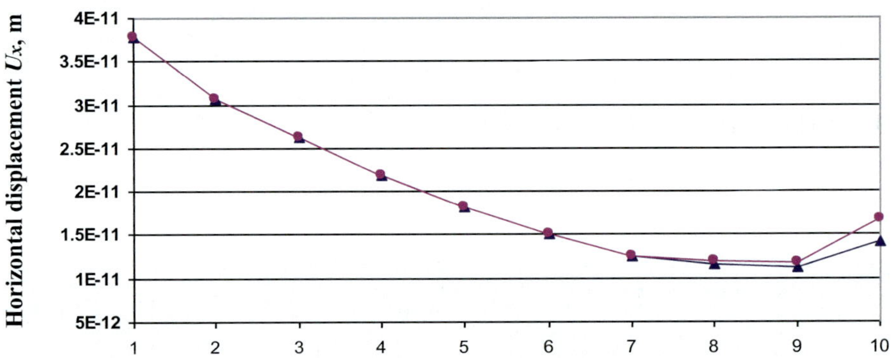

Fig. 9.52 Plots of the maximum values of displacements U_x at points on the section *A* (5, 0)–*B* (50, 0) of the anomalous structure "soft–hard–soft" and dots correspond to the case of an inclination angle of 45° and triangles correspond to the case of an inclination angle of 26.5°

When waves propagate in the region near the surface of the middle, more rigid layer, the oscillation amplitude decreases. With the subsequent propagation of vibrations to the right (after passing the slope), their level is significantly lower than in the surface layer to the left of the load source.

Case 2: Load to the Right of the Slope

The most characteristic distributions of displacement amplitudes at different times (when a wave packet passes through a slope) are presented in Figs. 9.54, 9.57, and 9.58. In this case (Fig. 9.2, case 2), the oscillation amplitude near the load source is higher than when it is located on a soft upper layer.

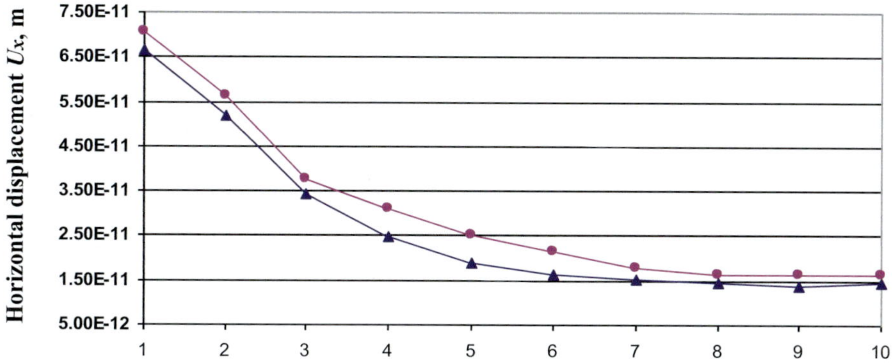

Fig. 9.53 Plots of the maximum values of displacements $|\vec{U}|$ at points on the section A (5, 0)–B (50, 0) of the anomalous structure "soft–hard–soft" and triangles correspond to the case of $h_1 = h_2 = 5$ m and dots correspond to the case of $h_1 = 10$ m and $h_2 = 5$ m

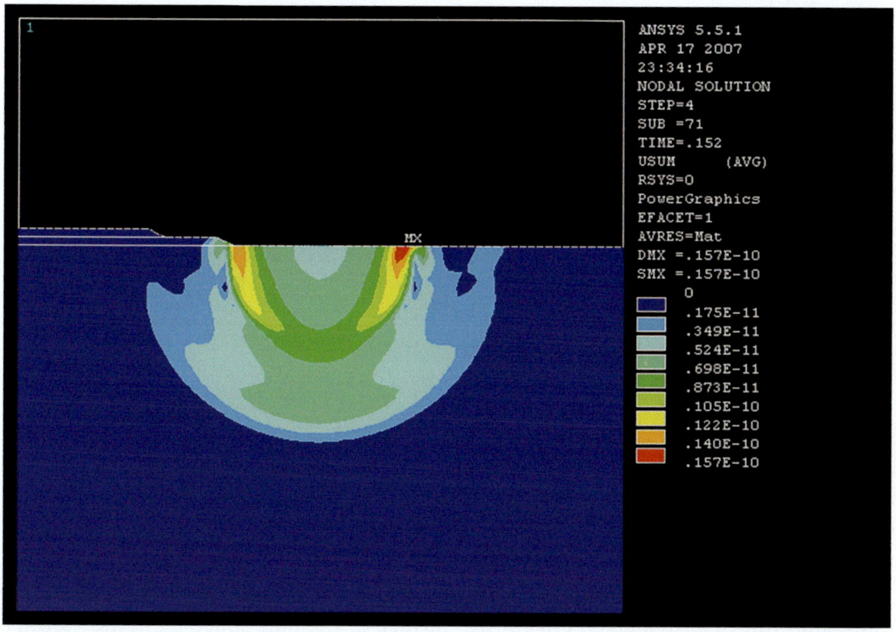

Fig. 9.54 Distribution of the displacement amplitudes at the time when the wave packet attains the lower face of the middle layer of slope ($t = 0.152$ s)

When the packet of waves attains the region of the lower surface of the second, hard layer, near the slope, a decrease in the displacement amplitude is observed (Fig. 9.54). With an anomalous "soft–hard–soft" structure, this decrease is significantly more than in other structures.

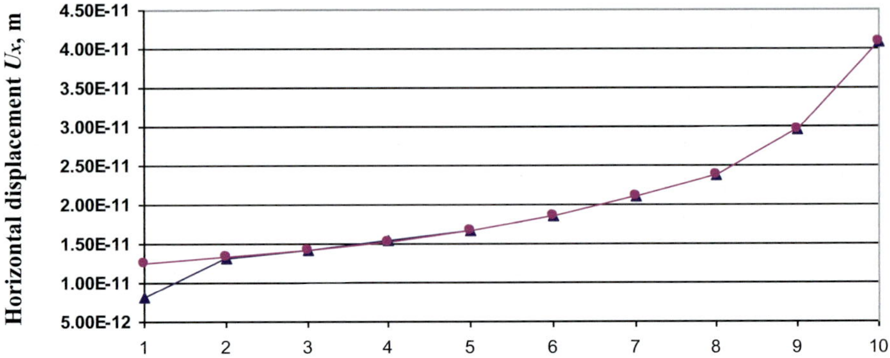

Fig. 9.55 Plots of the maximum amplitudes of displacements U_x at points on the section E (−50, 0)–F (−5, 0); dots correspond to the anomalous structure "soft–hard–soft" without a slope and triangles correspond to the anomalous structure "soft–hard–soft" containing the slope

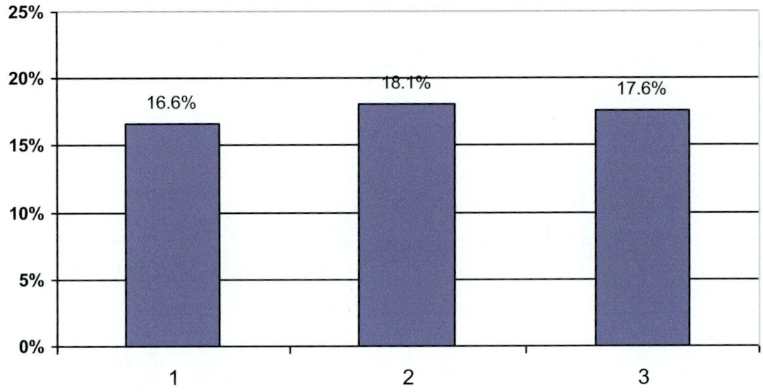

Fig. 9.56 Histograms of the percentage increase in the maximum displacement amplitudes $|\vec{U}|$ due to the inclination at point E of the structure of the anomalous structure "soft–hard–soft": (1) at inclination angle of 26.5° and $h_1 = h_2 = 5$ m; (2) at inclination angle of 45° and $h_1 = h_2 = 5$ m; (3) at inclination angle of 26.5° and $h_1 = 5$ m and $h_2 = 10$ m

A quantitative assessment of the influence of the slope on the change in the maximum amplitudes of displacements of the surface of the half-space U_x in the vicinity of the lower part of the slope is shown in Fig. 9.55.

Changes in the angle of inclination and the thickness of the middle layer have little effect on changes in the amplitude of oscillations of the surface of the half-space near the bottom of the slope. Figure 9.56 shows histograms of the percentage decrease in the maximum displacement amplitudes from the influence of the angle of inclination at point E for various geometric characteristics of the structure and slope layers.

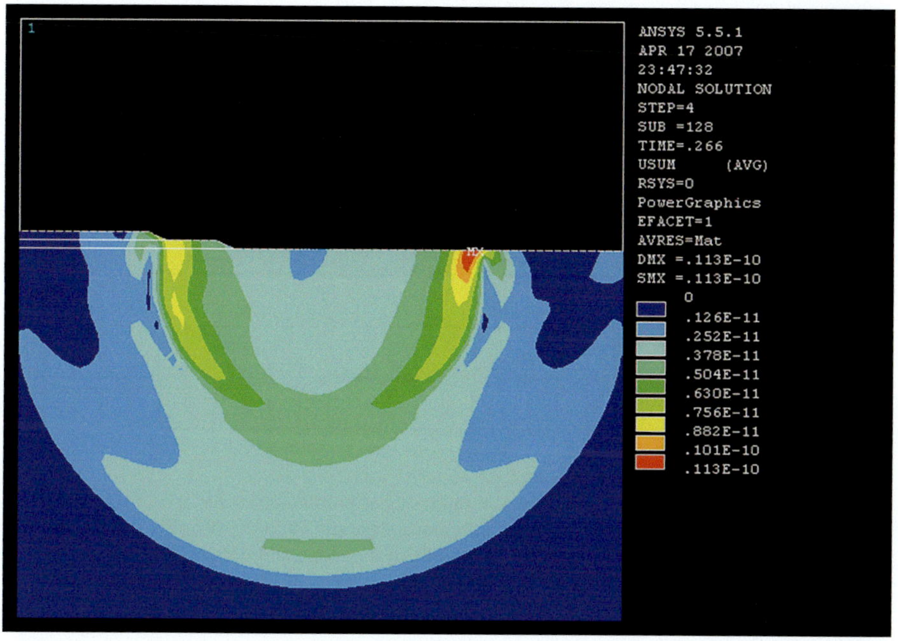

Fig. 9.57 Distribution of the displacement amplitudes at the time of propagation of the wave packet in the surface region of the first layer of slope ($t = 0.266$ s)

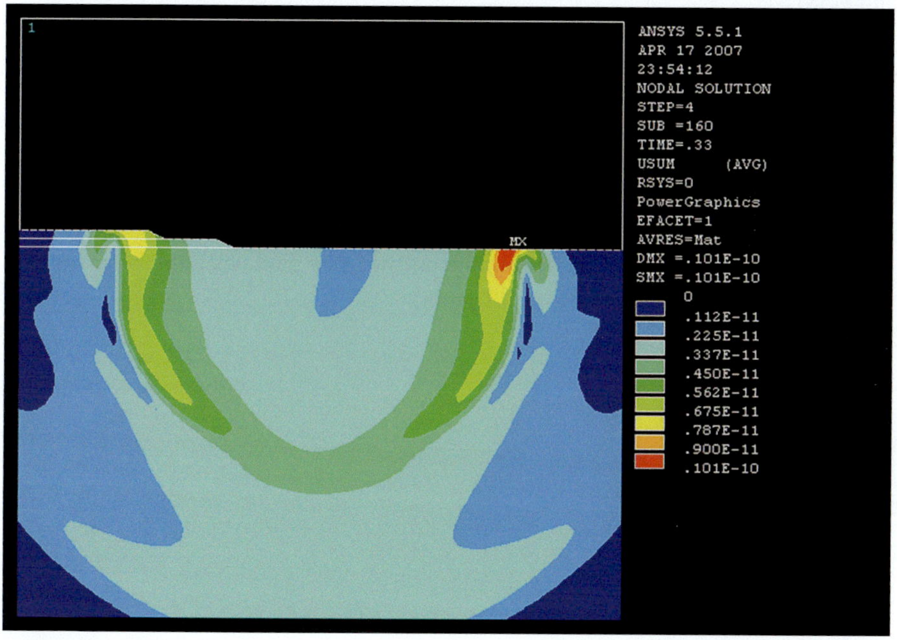

Fig. 9.58 Distribution of the displacement amplitudes at the time of propagation of the wave packet in the surface region of the first layer of slope ($t = 0.33$ s)

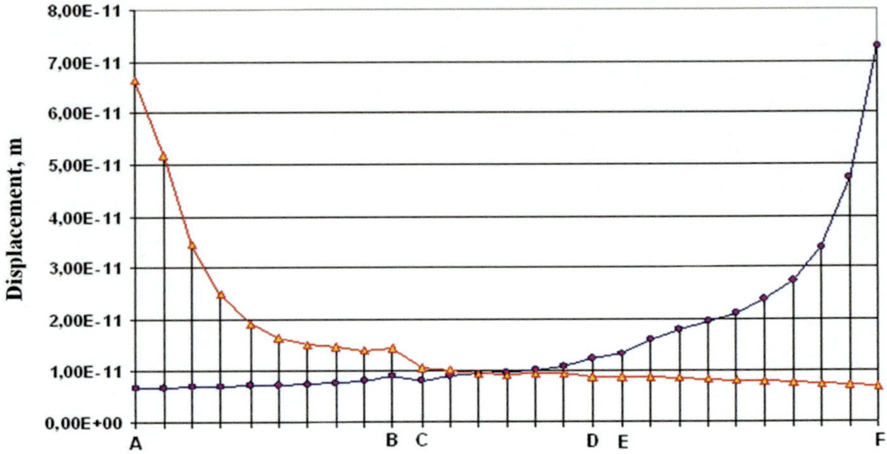

Fig. 9.59 Plots of maximum displacement amplitudes at points on the surface of the first (*AB*), second (*CD*), and third layers (*EF*) of the anomalous structure "soft–hard–soft"; triangles correspond to the case when loading is located to the left of the slope and dots correspond to the case when loading is located to the right of the slope

At the further propagation of oscillations along the slope and behind it to the left, a significant decrease in the amplitude of oscillations is observed (see Figs. 9.57 and 9.58).

Changes in the maximum displacement amplitudes $\left|\vec{U}\right|$ on the surface of various slope layers in the anomalous structure "soft–hard–soft" and their dependence on the linear coordinate over surface sections along the broken line *ABCDEF* at different positions of the vibration source relative to the slope are shown in Fig. 9.59.

An analysis of the results of the numerical experiments illustrates that when the source of man-made or seismic vibrations is located to the left of the slope of the anomalous structure "soft–hard–soft," the value of possible impact on objects located above the slope and on it is significantly higher than when the waves propagate from the source located below the slope.

Chapter 10
Modeling Pavement Constructions

Abstract This chapter presents the results of a study of the dynamic deformation characteristics of elements of layered extended structures using the example of the "road construction-soil" system. The FEM model developed and the results obtained on its base are used to modernize methods for predicting the process of damage accumulation in the elements of road constructions. For modeling semi-infinite media, the plane, spatial, and symmetric models are used. Deflections of the surface of coating are studied with dependence on geometric, physical, and mechanical characteristics of the considered layered structures.

Keywords Pavement construction · Road structure—soil system · Step method for non-stationary problem · Batch mode · Geometric and mechanical characteristics · Frequency response · Spatial and symmetric models · Bending stiffness

Chapter 9 presents the results associated with solving the problems of dynamic deformation of layered structural elements by finite-element methods. In the practical calculation of the amplitude displacement functions under a monoharmonic action based on the obtained integral representations, it seems possible to obtain stable results for a large number of system layers. However, this requires rather large resources, especially when calculating the modes of nonstationary influence, since it is necessary to take into account a large number of harmonics when using the harmonic analysis method. Moreover, it should be noted that in the calculation of stresses, the convergence of the corresponding improper integrals is significantly worse than in the calculation of displacements. As a result, there is a technical restriction of the possibilities of calculating stresses from analytical representations. For this reason, it is necessary to develop a finite-element model of a semi-infinite structure using damping belt technology when replacing an infinite (or very extended) volume with a finite one. To verify the reliability of this model, one can use the results of calculations of the amplitude displacement functions from analytical representations. When modeling such a semi-limited system as "road construction-soil," it is natural to use a large amount of

A. A. Lyapin et al., *Improving Road Pavement Characteristics*, Innovation and Discovery in Russian Science and Engineering,
https://doi.org/10.1007/978-3-030-59230-1_10

actual data obtained on the base of full-scale experimental studies using a mobile vibration measuring complex developed at the DorTransNII of the Rostov State Building University. As a result of such a two-level verification, it is possible to control not only the reliability of the results obtained, but also the adequacy and correspondence of the calculated model to reality.

This chapter presents the results of a study of the dynamic deformation characteristics of elements of layered extended structures using the example of the “road construction-soil” system. The FEM model developed and the results obtained on its base were used to develop methods for predicting the process of damage accumulation in the elements of road structures.

10.1 Statement of Problems into Framework of FEM

To obtain the correct results for any simulation, it is important to get as close as possible to the system of representations, dependencies, conditions, and restrictions that describe the process or phenomenon being studied and calculated. In other words, it is necessary that the model describes not only the geometry of the real structure and the ratio of the mechanical characteristics of its elements, but also the features of the dynamic effect on it during testing and its operation.

The construction of the computational model is always based on a number of simplifying assumptions related to both the idealization of the defining relations and the inevitable error in setting the mechanical characteristics of its elements and external operating loads. Therefore, any model in the development process has a “fatal error,” the value of which is determined during the verification of its adequacy to a real object.

To study the characteristics of the dynamic deformation of layered structures, a model of the multilayer system of “road structure-soil” was created. This model was developed using the numerical method, namely the finite element method (FEM), using the ANSYS software package. The base for the formulation of model problems of mechanics that correctly describe the features of loading and deformation of the elements of the multilayer system of “road structure-soil” is the geometry and structure of real road structures.

The system of “road structure-soil” (see Fig. 10.1) includes the coating layers, the base layers, the soil of the subgrade (working layer), and the soil massif on which the road structure is located. The design of pavement is described by a set of interconnected strips with plane-parallel borders. Each element of the calculation model is characterized by certain physical and mechanical properties, such as elastic modulus (E), density (ρ), Poisson’s ratio (ν), and viscosity (η). Road structures in straight sections are limited in width, but have a sufficient length.

The impact on the structure during operation is determined by transport moving along its surface and is nonstationary dynamic with a fairly wide frequency spectrum. On the other hand, when testing road constructions, shock impact setups with a nonstationary nature of impact into a wide frequency range are most often used.

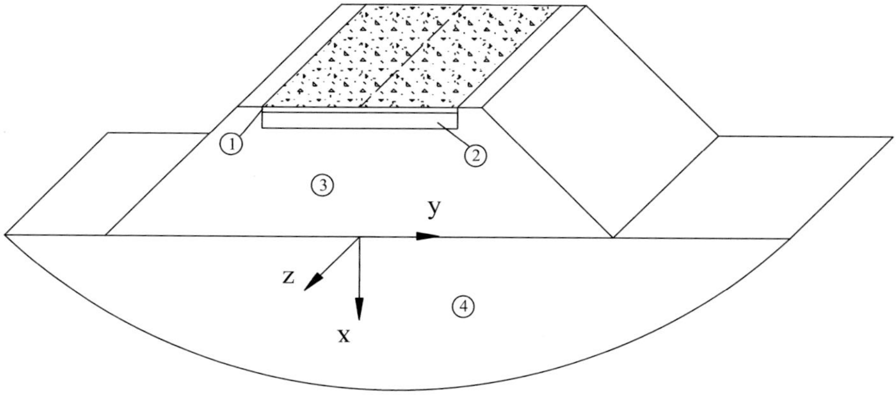

Fig. 10.1 System of "road construction-soil": 1—coating; 2—base; 3—soil subgrade; 4—underlying soil massif

This determines the feasibility of applying nonstationary analysis methods in the development of the FEM model.

By solving the problem, the method of "continuation with respect to the time parameter" or the so-called step method for solving the nonstationary problem is used. The time segment is divided into steps, the length of which is determined by refining the system based on the accuracy requirements of the result.

We also note that the determining factors affecting the dynamic process of generation and propagation of stress and strain fields in system elements are the mechanical properties of materials and the structure of the road construction. Therefore, it is obvious that the task of determining the stress-strain state of the road construction under dynamic action requires iterative calculations, with adjustment of various design parameters. In this regard, this calculation model is implemented in the so-called batch mode. This method allows one to change the original text of the batch file, which determines both the geometric and mechanical characteristics of the system elements.

It is known that the frequency response of an impact load is determined by the amplitude-time response of a change in the resulting contact interaction of colliding bodies $\vec{P}(t)$. Moreover, the shorter the collision time τ, the wider the frequency spectrum $\vec{P}(\omega)$ of the impact and the closer to the main value (amplitude) its main part (see Fig. 10.2). Under the influence of force systems on the coating surface, the frequency spectrum is determined by the Fourier transform [15, 131] of the function $\vec{P}(t)$:

$$\vec{P}(\omega) = \int_{-\infty}^{+\infty} \vec{P}(t) \cdot \exp(-i\omega t)\,\mathrm{d}t \tag{10.1}$$

where ω (rad/s) is the circular oscillation frequency.

To assess the mechanical characteristics of the road construction elements, a hard impact scheme was selected for a stamp in contact with the surface of the

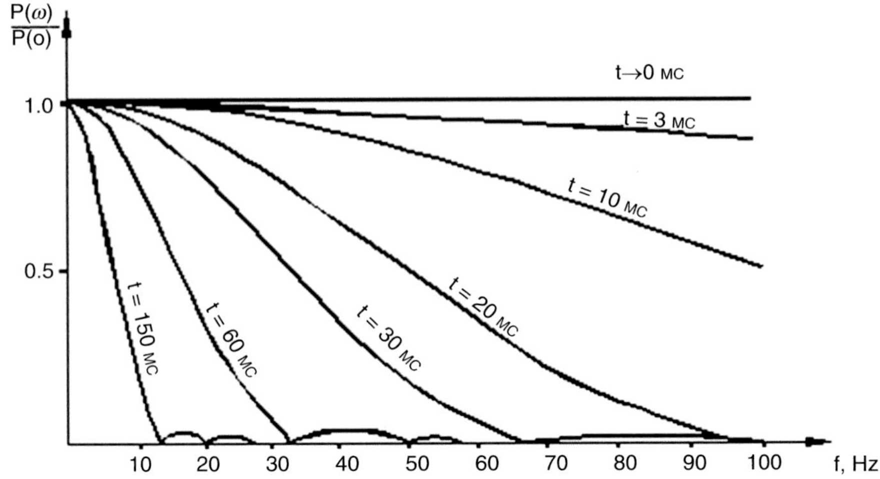

Fig. 10.2 Dependence of the frequency response of impact on the pulse duration

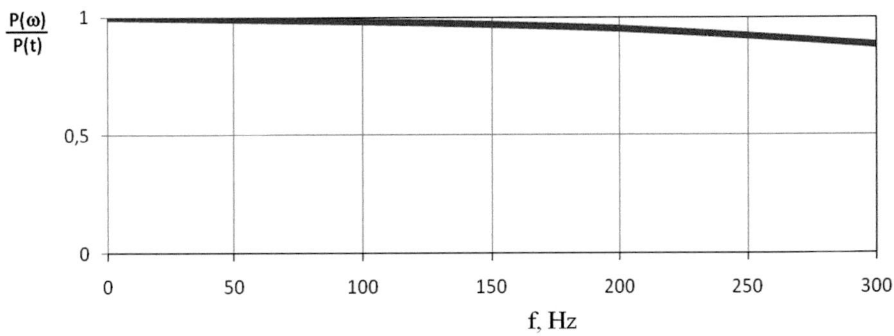

Fig. 10.3 Frequency response of a triangular pulse in the frequency range of 0–300 Hz

asphalt concrete pavement with an impact duration of 3 ms, which allows one to obtain a fairly broadband spectrum of influence in the frequency range up to 300 Hz with an equilibrium energy distribution over all frequency components (Fig. 10.3).

In shape, in experiment, the impact is modeled by a triangular pulse of the form shown in Fig. 10.4. The pulse amplitude is normalized so that the spectral function is close to unity (with this normalization, the pulse characteristic has a frequency dimension). It is assumed that the distribution of forces along the contact patch (circle of radius 0.125 m) is constant.

To study the patterns of deformation of a road construction upon impact and subsequent propagation of oscillations and waves in its elements, plane and spatial models have been developed and studied. Plane models (corresponding to the longitudinal or transverse section of the structure) can be used to study the process in the corresponding section at relatively short time ranges after impact. These models require additional normalization of the maximum surface pressure upon impact.

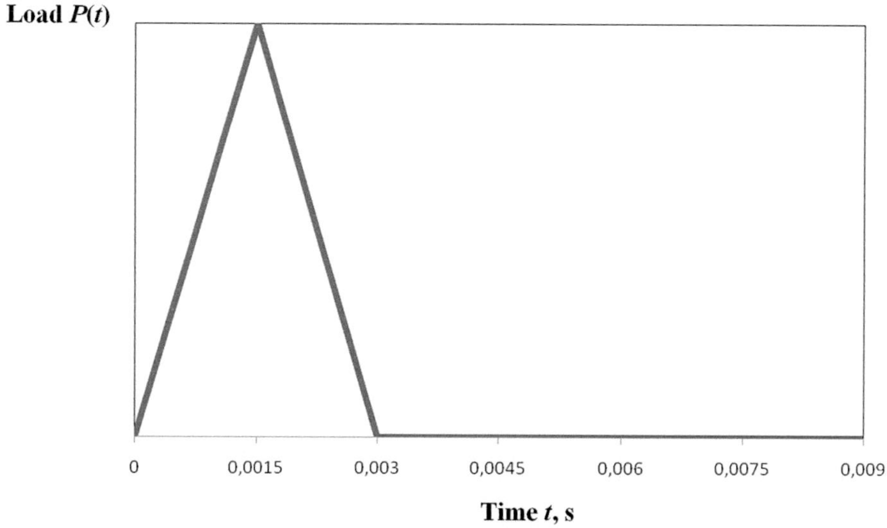

Fig. 10.4 Amplitude-time characteristic of the impact pulse force

The calculation time for these models, compared with the spatial model, is minimal. The spatial model is complex, and does not have the limitations inherent to plane models, but requires large resources to obtain each result. For this reason, it is advisable to use it to analyze cases in which the use of simplified models is incorrect or leads to large calculation errors.

10.2 Features of Modeling Semi-infinite Media

The mechanisms of wave field propagation in layered structures, including road constructions, were considered in [4, 17, 66, 91, 121, 129]. So, if the medium filling the half-space is divided into layers, that is, each layer of a given thickness is characterized by certain properties (elastic modulus and density), then under the action of short-term loads on its surface, a plane compression wave is generated in the upper layer (vector 1 in Fig. 10.5), which, meeting the layer boundary, is converted into reflected and refracted waves (vectors 2 and 3 in Fig. 10.5).

As a result of repeated reflections and transformations in a layered medium under dynamic action, the following types of waves are observed:

(i) *Surface waves*, such as Rayleigh waves, propagating along the free (day) surface of structural elements whose amplitude decays exponentially with distance from the surface: The rate of damping of oscillations with distance from the area of loading to the depth of the medium corresponds to the law

$$A(r) = \frac{A \cdot \exp(-\delta \cdot r)}{\sqrt{r}}$$

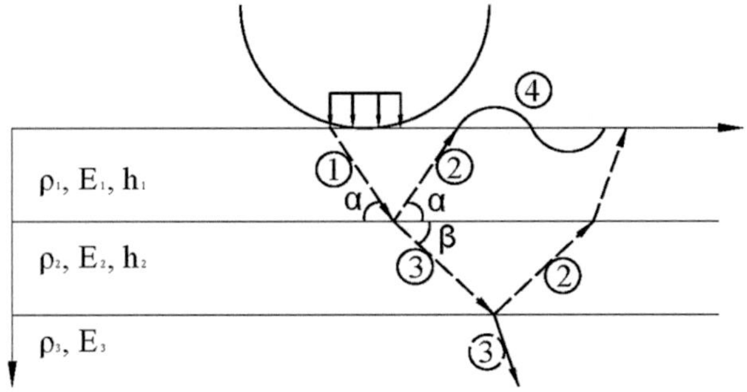

Fig. 10.5 Wave processes in the multilayer system of "road construction-soil"

where $A(r)$ is the amplitude of the oscillations at a distance from the source r; r is the distance from the source of oscillations to the point of the medium; and δ is a damping coefficient of oscillations.

(ii) *Boundary (channel) waves* propagate along the interfaces of the layers, and decrease exponentially with distance from the boundaries. The law of decreasing oscillation amplitude is the same as for a surface wave.

The propagation velocity of surface and boundary waves is determined by the material density ρ and its elastic properties (λ and μ are Lamé coefficients), and is close to the propagation velocity of shear waves V_s.

(iii) *Internal waves*: Longitudinal, with propagation speed V_p, and transverse, with propagation speed V_s.

The following general patterns have been identified. All types of waves in the elements of the system interact with each other, and transform when passing through the interface of the layers and when reflected from the side surfaces of the structural layers, which determines a very complex picture in the zone closest to the area of the loading of the structure. Part of the energy of the wave field, passing through the structural layers of the pavement, is transmitted to the soil of the subgrade and the underlying soil massif. The other part in the process of re-reflection, refraction, superposition, and dispersion again reaches the surface of the coating, and at the same time undergoes changes due to the geometric parameters of the pavement and the physical properties of its elements.

By using the FEM, certain difficulties arise associated with modeling the infinity of the system of "road structure-soil," due to the type and structure of the construction:

(i) Choice of representative volume
(ii) Determination of the optimal partition of the representative volume into finite elements, providing the required accuracy of the final result while minimizing the estimated time
(iii) The introduction of peripheral damping belts to eliminate the influence of reflected waves from the boundaries of the region when it is approximated as a finite body

The initial size of the representative volume in the study of nonstationary effects on the system (for example, impact load) was determined by the time of propagation of the longitudinal wave with the speed V_p along the shortest path to the nearest boundary of the representative volume and back to the observation point.

To correctly select the size and shape of the representative volume, we used a comparison of the frequency characteristics of the vibration of surface points in the central part of the structure with those calculated by the analytical axisymmetric model.

For each type of system model, a decrease in the maximum size of a finite element leads to an increase in the dimension of a set of linear algebraic equations (LAEs) obtained as a result of discretization of the problem. The same effect is observed with an increase in the size of the representative volume.

With an increase in the dimension of the LAEs set, a rapid increase in the calculation time is observed and, from some critical value of the dimension, a decrease in the accuracy of the result occurs due to the accumulated calculation error. For this reason, in developing this finite-element model, much attention was paid to the choice of the sizes of the representative volume and its partition into finite elements.

Let us dwell in more detail on the features of testing computational models that provide reliability and the required accuracy of the calculation. The calculations are carried out, the main purpose of which is to verify the correct choice of the lattice for partitioning the region into finite elements (FEs) and the sizes of the representative volume and the properties of the damping belts. The choice of the final maximum size of the finite elements was determined by the sequential calculation of the task for an increasingly diminishing lattice size. The required accuracy was determined by a relative error not exceeding 3%. According to the calculation results for various sizes of the representative volume, it is gradually seen how the oscillations propagating in the structure approach the fictitious boundary of the representative volume, pass through damping belts, are reflected, and begin to propagate in the opposite direction (to the source of oscillations).

If it is necessary to reduce the calculation time, a rougher model can be used, the accuracy of the calculation results for which is already known from the numerical experiment. The presented scheme of choosing the representative volume of the model and dividing it into FEs is valid for both plane and spatial setting.

When choosing the size of the region and the viscosity of the damping belts, it was assumed that the most intense vibrations arriving at the observation point after reflection from the fictitious boundary should not exceed 0.01 of the amplitude of the direct vibrations arriving at the observation point from the loading source.

As an example, Fig. 10.6 shows dependence of the amplitudes of vertical oscillations on time at the observation point located in the longitudinal section of the structure at a distance of 1.0 m from the side edge of the coating and near the impact load.

The most critical stage is the partitioning of the model geometry into a finite-element lattice, since the accuracy of the future solution and the calculation time depend on the number of finite elements. It is almost always advisable to use a lattice with a thickening in that part of the region where a faster change in the design

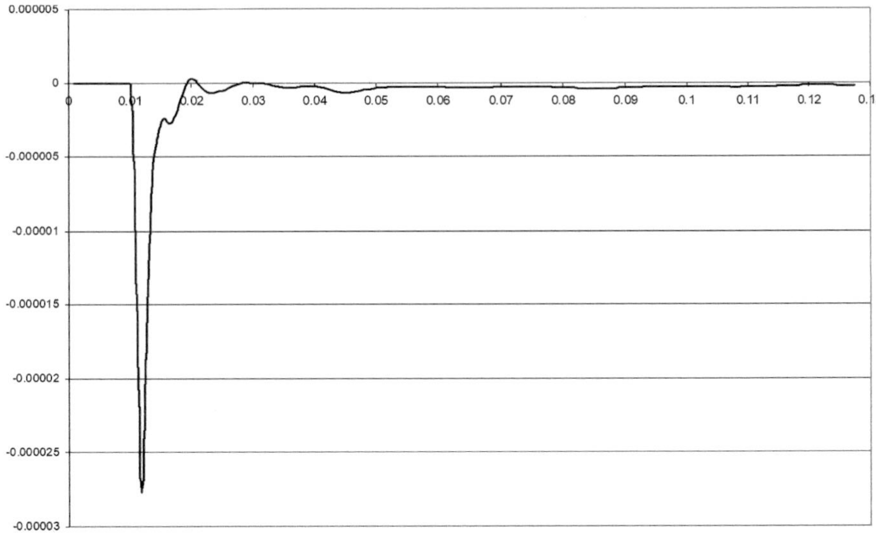

Fig. 10.6 Plot of amplitudes of vertical oscillations (spatial model)

characteristics is observed. In this case, mesh thickening should be carried out in the vicinity of the loading region and near the boundaries of subdomains with significantly different physical and mechanical characteristics of the material.

The thickening is carried out not for the entire lattice as a whole, but in separate fragments in which the behavior of the solution substantially depends on the degree of discretization. In our tasks it is:

(i) The area in the vicinity of which the load is applied
(ii) The boundaries of zones with different physical, mechanical, and geometric characteristics
(iii) Areas of direct interest for identifying the main relationships between the stress-strain state of the system and the properties of its individual elements

In order to suppress the severity of the intrinsic resonances of the system of "road construction-soil," as a massive object, it is necessary to have sufficiently large linear dimensions of the model compared to the width of the pavement (this will exclude the possibility of re-reflection of waves from side faces). The practical implementation of the problem under consideration by the finite element method eliminates the possibility of modeling the infinite geological medium. Therefore, a damping layer of considerable thickness, having mechanical properties corresponding to the structure of the geological medium, is introduced along the contour of the construction.

The accuracy of the solution in the area of the damping layer is not critical. The function of the damping layer, as noted above, is to suppress the intrinsic low-frequency resonances of the system, and to eliminate the possibility of the influence of reflected waves on the accuracy of solving the entire problem.

In connection with the above conditions, it has the meaning to gradually reduce the accuracy of the partition at moving away from the impact site. With such a partition, the accuracy of the solution in the zone closest to the place of impact does not decrease and the calculation time is significantly reduced. As an example of such a partitioning scheme, Fig. 10.7 shows a finite-element lattice.

10.2.1 Spatial Model

The spatial model includes the selection of a limited part of the volume occupied by the road construction and the underlying soil centered at the point where the impact is applied to the surface of the system. We obtain the initial data for choosing the geometric dimensions of this region based on that the observation time at a given point t is significantly shorter than the time range t_1 ($t << t_1 = l_{min} / V_{pmax}$) necessary for the longitudinal wave propagating in the layered structure (with velocity V_{pmax}) to travel from the source of vibrations to the nearest fictitious boundary of the representative volume and back to the observation point (total distance l_{min}). To increase the reliability of the results, two belts of damping elements are created along the perimeter of the region, which have the same mechanical characteristics as those located nearby inside the representative volume, but have increasing viscosity ($\eta_1 = 0.01$, $\eta_2 = 0.02$).

An example of dividing the spatial representative volume of the system of "road construction-soil" into finite elements, including damping belts, is presented in Fig. 10.8.

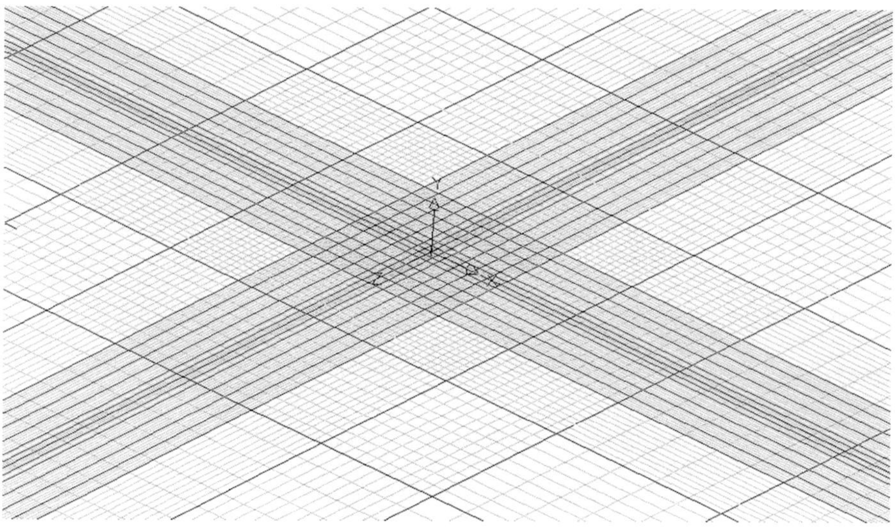

Fig. 10.7 Fragment of the finite-element lattice of the studied multilayer system of "road construction-soil"

As examples, Fig. 10.9 shows distributions of vertical displacements at times $t_1 = 0.015$ s and $t_2 = 0.024$ s.

The spatial model of the system is rather cumbersome in calculations and requires a large amount of technical work when changing it. The high order of the system of linear algebraic equations, to which it reduces, also negatively affects the obtained accuracy of the result. This determines the feasibility of developing a simplified symmetric model.

10.2.2 Symmetric Model

Assume that the impact load is applied at the point of the center line of the coating (or in the immediate vicinity of it), which allows considering only its quarter due to symmetry of the structure (Fig. 10.10). This model can significantly reduce the complexity of generating a finite-element lattice and reduce the order of the LAE systems being solved (and, accordingly, the calculation time).

Comparative calculations using spatial and symmetric models showed that when the impact point deviates from the center line by no more than 10% of the coating width, the results of calculations using the symmetric model give satisfactory

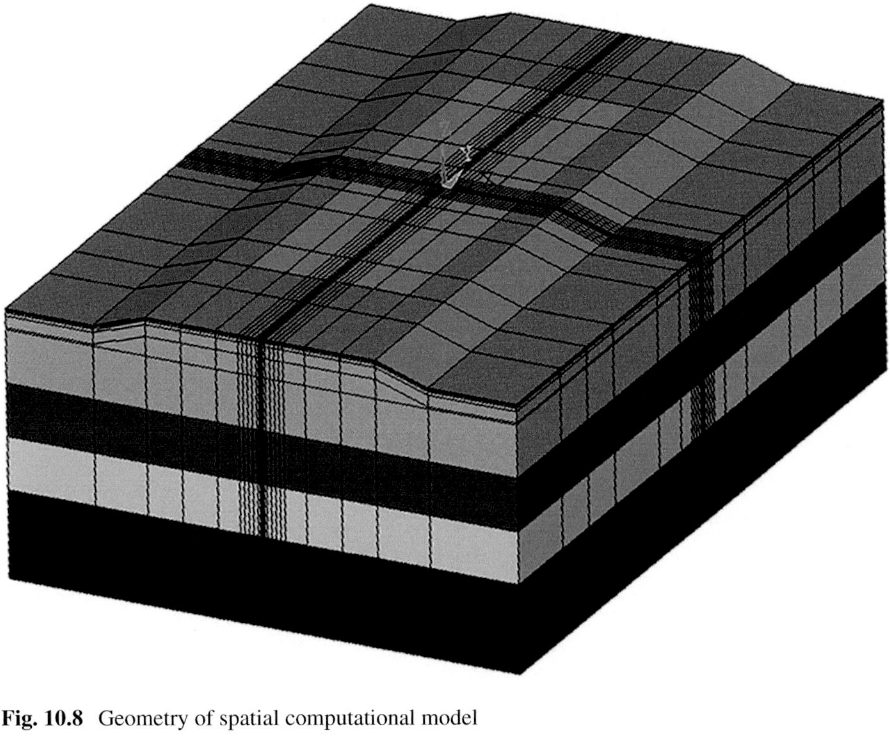

Fig. 10.8 Geometry of spatial computational model

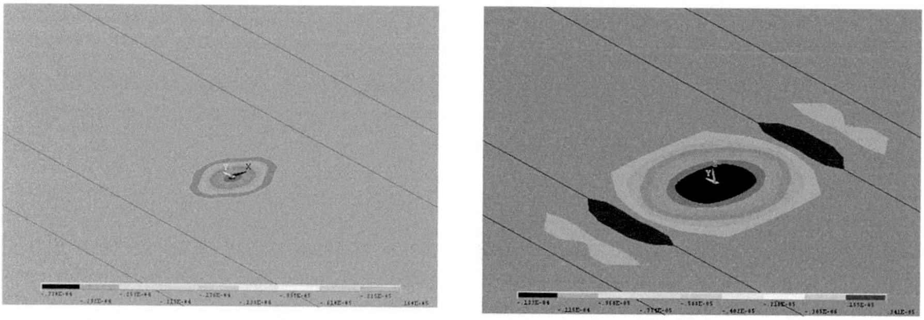

Fig. 10.9 Distributions of vertical displacements at times $t_1 = 0.015$ s and $t_2 = 0.024$ s

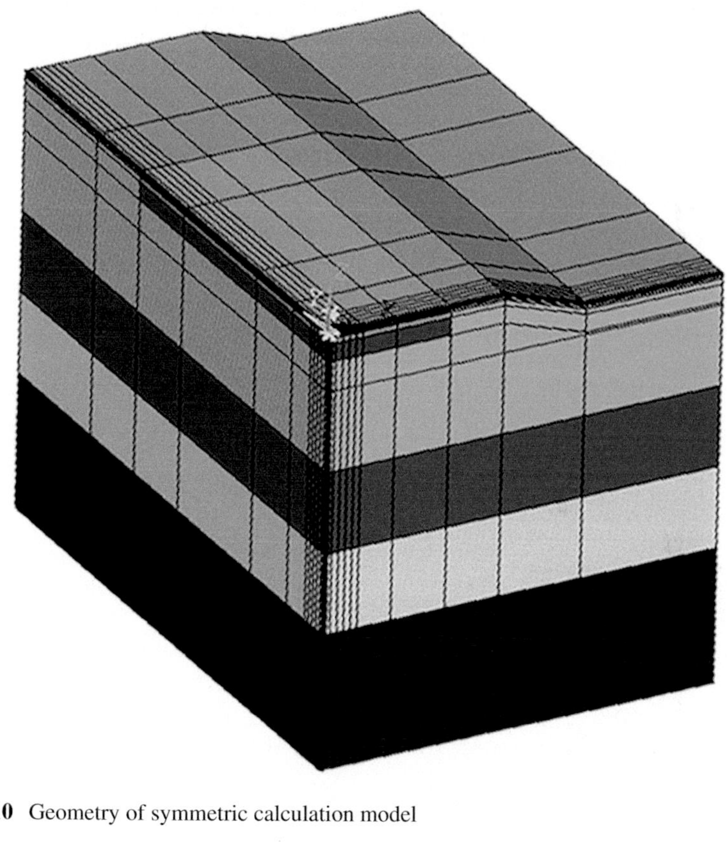

Fig. 10.10 Geometry of symmetric calculation model

accuracy (of the order of 8–12%) in almost the entire time range, which coincides well with three-dimensional case at short times.

An additional criterion for the adequacy of the calculated model of reality is a good agreement between the calculated data and the materials of the full-scale experiment conducted on roads, the designs of which are known and implemented in the mathematical model (see Fig. 10.11).

10.2.3 Plane Model

Plane models corresponding to the longitudinal or transverse cross section of the structure can be used to study the process in the corresponding section at relatively short ranges after impact. These models require additional normalization of the maximum surface pressure upon impact. The calculation time for these models, in comparison with the symmetric and spatial models, is minimal.

As an example, Fig. 10.12 shows the structure of the representative volume, its partition into finite elements, and damping belts for a plane model.

10.3 Comparative Analysis of Characteristics

Based on the developed models, detailed numerical experiments were carried out aimed at studying the influence of the structure of the road construction and the state of its elements on the characteristics of the dynamic stress-strain state of the coating surface and structural elements.

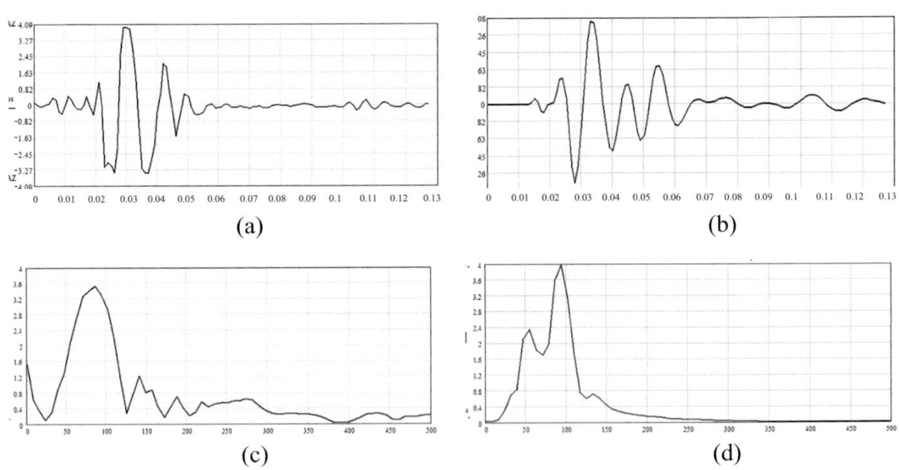

Fig. 10.11 Amplitude-time characteristic of the known road construction: (**a**) distance from the center of impact 1.25 m (experimental data); (**b**) distance from the center of impact 1.25 m (calculation results) and amplitude-frequency characteristic of the same road construction; (**c**) distance from the center of impact 1.25 m (experimental data); (**d**) distance from the center of impact 1.25 m (calculation results)

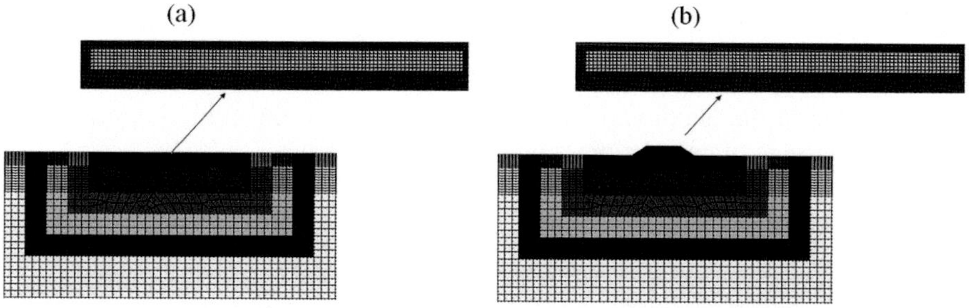

Fig. 10.12 Structure of a plane model with a partition into finite elements: (**a**) longitudinal cross section, (**b**) cross section

The *first direction* of the numerical experiments is aimed at revealing the information content of various characteristics of the surface response when the state of structural layers changes. For this purpose, one type of road construction was considered, in which the mechanical properties of the coating, base, and soil of the subgrade were varied. The influence of the mechanical characteristics of the structural layers (reflecting the state of this element) on the formation of instant cups and cups of maximum deflections of the surface, as well as on the values of maximum tensile forces near the upper and lower boundaries of the asphalt concrete layers, is studied.

The *second direction* of the numerical experiments is based on calculations of the dynamic stress-strain state characteristics for two types of road construction of the same width. The road construction of enhanced strength (with a thickness of asphalt concrete layers of more than 18 cm) and "medium" strength (with a thickness of asphalt concrete layers of 10–12 cm) were selected. Calculation of cups of instant and maximum deflections was carried out not only for the coating surface, but also for horizontal sections of the construction. The latter is necessary to identify the contribution of deformation of structural elements to the complete deformation of the surface of the construction.

10.3.1 Initial Data of Numerical Experiment

During the numerical experiment, the road construction was considered as a three-layer system: coating layers (monolithic layers of asphalt concrete with bending stiffness), base layers (disconnected layers), and subgrade. In this case, the elastic moduli of the coating and the base were calculated as the weighted average values of several layers. The ground massif was modeled as a two-layer half-space (a layer on a more rigid half-space). The calculated characteristics of the simulated road constructions are presented in Table 10.1.

The calculations were carried out for surface points and internal points of the structure along its longitudinal, transverse, and horizontal sections. Based on the calculation results, plots were constructed of instant cups of surface deflections or horizontal sections of the road construction, as well as cups of maximum deflections. The latest plots were based on the choice at each point of the surface or horizontal section of the maximum vertical displacement for the entire period of time during the passage of calculations from the point of impact.

10.3.2 Response of the Surface of Coating to Impact Loading for Different Mechanical Properties of Road Construction Elements

The road structure No. 1 (Table 10.1) with a common elastic modulus of 390 MPa was adopted as the base one. In construction Nos. 1, 2, 3, 1(1), 1(2), 1(3), 2(1), 2(2), 3(1), and 3(2) (Table 10.1), only mechanical characteristics of the coating layers, base, and soil of the subgrade were varied. Based on the calculation results, cups of instant and maximum dynamic surface deflections were built, and the maximum tensile forces were calculated near the upper and lower boundaries of the asphalt concrete layers. The most characteristic plots reflecting the nature of the change in the deflections formed upon the impact of the cups are given in Figs. 10.13, 10.14, 10.15, 10.16, 10.17, 10.18, 10.19, 10.20, 10.21, 10.22, 10.23, 10.24, 10.25, 10.26, 10.27, and 10.28. All necessary explanations are made in the figure captions.

Based on the analysis of the patterns obtained as a result of systematization of the data of a numerical experiments, the following conclusions can be drawn:

1. The amplitude and geometric characteristics of the cups of instantaneous deflections of the surface of the coating noticeably change when the mechanical characteristics of the asphalt concrete layer change. Changes in the mechanical characteristics of other structural elements are determined by less significant and poorly reflected in the plot changes in the cups of instant deflections for a solid structure (Figs. 10.13, 10.15, and 10.17).
2. The cups of maximum deflections to a greater extent reflect the change in the mechanical characteristics of all structural layers; that is, this characteristic is more informative (from the point of view of assessing the state of structural layers) than the cup of instant deflections (Figs. 10.14, 10.16, and 10.18).
3. A decrease in the elastic modulus of the coating (asphalt concrete layers) leads to a significant increase in the dynamic deflection in the zone closest to the impact, namely at a distance of up to 0.5 m (Figs. 10.13 and 10.14).
4. The change in the mechanical characteristics of the base is manifested in a change of the dynamic deflections at a distance of 0.25–0.75 m from the point of impact (Figs. 10.15 and 10.16).

Table 10.1 Design characteristics of the elements of the model of "road construction-soil"

	Model elements "road construction-soil"		
Calculated construction parameters	Asphalt concrete (ac)	Base layers (bl)	Subgrade (s)
Construction no. 1 ($E_{com} = 390$)			
Elastic modulus (E), MPa	2300	283	47
Density (ρ), kg/m^3	2250	1900	1700
Poisson's ratio (ν)	0.35	0.3	0.35
Thickness (t), m	0.2	0.5	3.0
Construction no. 1(1) ($E_{com} = 373$)			
Elastic modulus (E), MPa	2000	283	47
Density (ρ), kg/m^3	2250	1900	1700
Poisson's ratio (ν)	0.35	0.3	0.35
Thickness (t), m	0.2	0.5	3.0
Construction no. 1(2) ($E_{com} = 342$)			
Elastic modulus (E), ML Pa	1500	283	47
Density (ρ), kg/m^3	2250	1900	1700
Poisson's ratio (ν)	0.35	0.3	0.35
Thickness (t), m	0.2	0.5	3.0
Construction no. 1(3) ($E_{com} = 301$)			
Elastic modulus (E), MPa	1000	283	47
Density (ρ), kg/m^3	2250	1900	1700
Poisson's ratio (ν)	0.35	0.3	0.35
Thickness (t), m	0.2	0.5	3.0
Construction no. 2 ($E_{com} = 289$)			
Elastic modulus (E), MPa	2300	283	20
Density (ρ), kg/m^3	2250	1900	1700
Poisson's ratio (ν)	0.35	0.3	0.35
Thickness (t), m	0.2	0.5	3.0
Construction no. 2(1) ($E_{com} = 463$)			
Elastic modulus (E), MPa	2300	283	90
Density (ρ), kg/m^3	2250	1900	1700
Poisson's ratio (ν)	0.35	0.3	0.35
Thickness (t), m	0.2	0.5	3.0
Construction no. 2(2) ($E_{com} = 217$)			
Elastic modulus (E), MPa	2300	283	10
Density (ρ), kg/m^3	2250	1900	1700
Poisson's ratio (ν)	0.35	0.3	0.35
Thickness (t), m	0.2	0.5	3.0
Construction no. 3 ($E_{com} = 267$)			
Elastic modulus (E), MPa	2300	120	47
Density (ρ), kg/m^3	2250	1900	1700
Poisson's ratio (ν)	0.35	0.3	0.35
Thickness (t), m	0.2	0.5	3.0
Construction no. 3(1) ($E_{com} = 339$)			

(continued)

Table 10.1 (continued)

Calculated construction parameters	Model elements "road construction-soil"		
	Asphalt concrete (ac)	Base layers (bl)	Subgrade (s)
Elastic modulus (E), MPa	2300	200	47
Density (ρ), kg/m^3	2250	1900	1700
Poisson's ratio (ν)	0.35	0.3	0.35
Thickness (t), m	0.2	0.5	3.0
Construction no. 3(2) (E_{com} = 307)			
Elastic modulus (E), MPa	2300	160	47
Density (ρ), kg/m^3	2250	1900	1700
Poisson's ratio (ν)	0.35	0.3	0.35
Thickness (t), m	0.2	0.5	3.0
Construction no. 4 (E_{com} = 168)			
Elastic modulus (E), MPa	2300	283	47
Density (ρ), kg/m^3	2250	1900	1700
Poisson's ratio (ν)	0.35	0.3	0.35
Thickness (t), m	0.1	0.2	3.0

5. A decrease in the modulus of elasticity of the soil in the cup of maximum dynamic deflections leads to an increase in the displacements in the far zone, namely at a distance of more than 0.75 m (Fig. 10.18).
6. An increase in the elastic modulus of asphalt concrete leads to an increase in tensile stresses for a strong structure and to an increase in tensile stresses at the lower and upper boundaries for weak structures (Figs. 10.19 and 10.20).
7. An increase in the elastic modulus of the base leads to a decrease in tensile stresses (Figs. 10.21 and 10.22).

10.3.3 Features of Deforming the Various Road Constructions

To study the basic laws of changing the amplitude and geometric characteristics of the cups of maximum deflections in the dependence of the structure of the road structure, numerical experiments were carried out, including the calculation of the dynamic stress-strain state characteristics of two types of road construction of the same width (construction Nos. 1 and 4, see Table 10.1).

In order to obtain the maximum change in the main characteristics of the stress-strain state, the road construction of enhanced strength (with a thickness of asphalt concrete layers of more than 18 cm) and medium strength (with a thickness of asphalt concrete layers of 10–12 cm) was selected. Calculation of instant cups (Figs. 10.23 and 10.25) and maximum deflections (Figs. 10.24 and 10.26) was carried out not only for the coating surface, but also for horizontal cross sections of the construction. The cups of maximum deflections to a greater extent reflect the change in the mechanical characteristics of all structural layers. Therefore, to analyze the laws of change in the characteristics of the dynamic deformation of road construc-

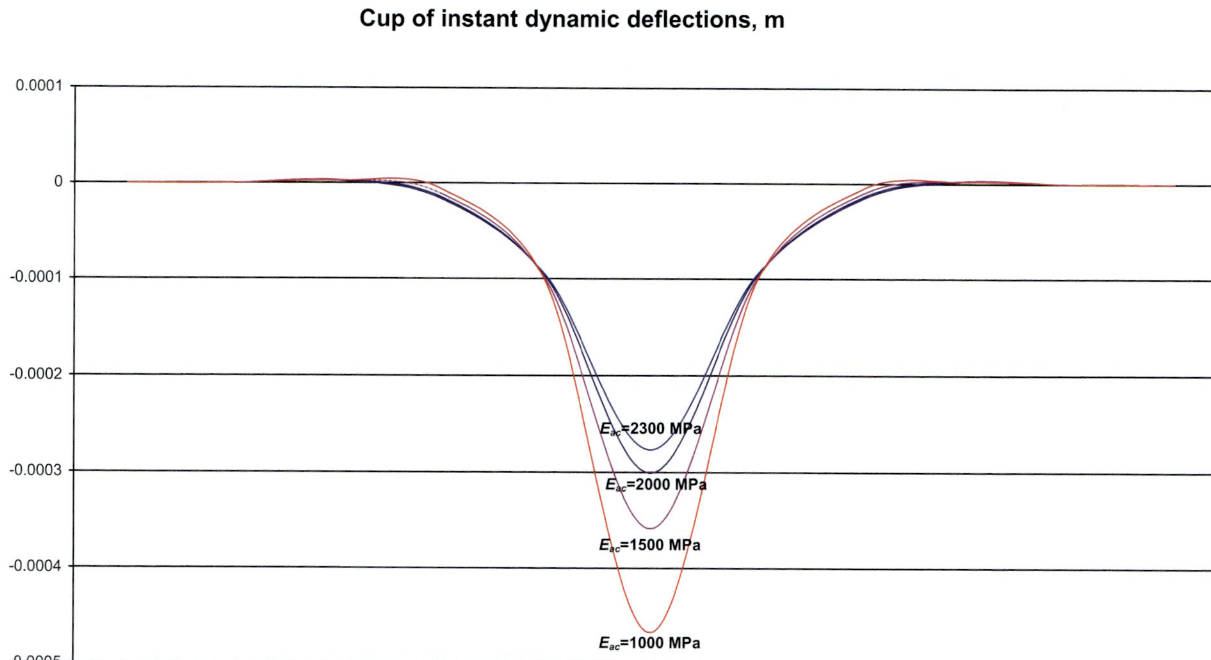

	2.5	1.25	0.75	0.5	0.25	0	0.25	0.5	0.75	1.25	2.5
Construction No.1	0.0000000040	-0.000000042	0.0000027039	-0.000016242	-0.000097211	-0.000276686	-0.000097211	-0.000016242	0.0000027039	-0.000000042	0.0000000040
Construction No.1(1)	0.0000000075	-0.000000048	0.0000029630	-0.000014545	-0.000099032	-0.000300621	-0.000099032	-0.000014545	0.0000029630	-0.000000048	0.0000000075
Construction No.1(2)	0.0000000056	-0.000000030	0.0000031003	-0.000010965	-0.000102151	-0.000358537	-0.000102151	-0.000010965	0.0000031003	-0.000000030	0.0000000056
Construction No.1(3)	0.0000000006	-0.000000003	0.0000025145	-0.000005973	-0.000104948	-0.000466691	-0.000104948	-0.000005973	0.0000025145	-0.000000003	0.0000000006

Fig. 10.13 Simulation of the calculated characteristics of coating

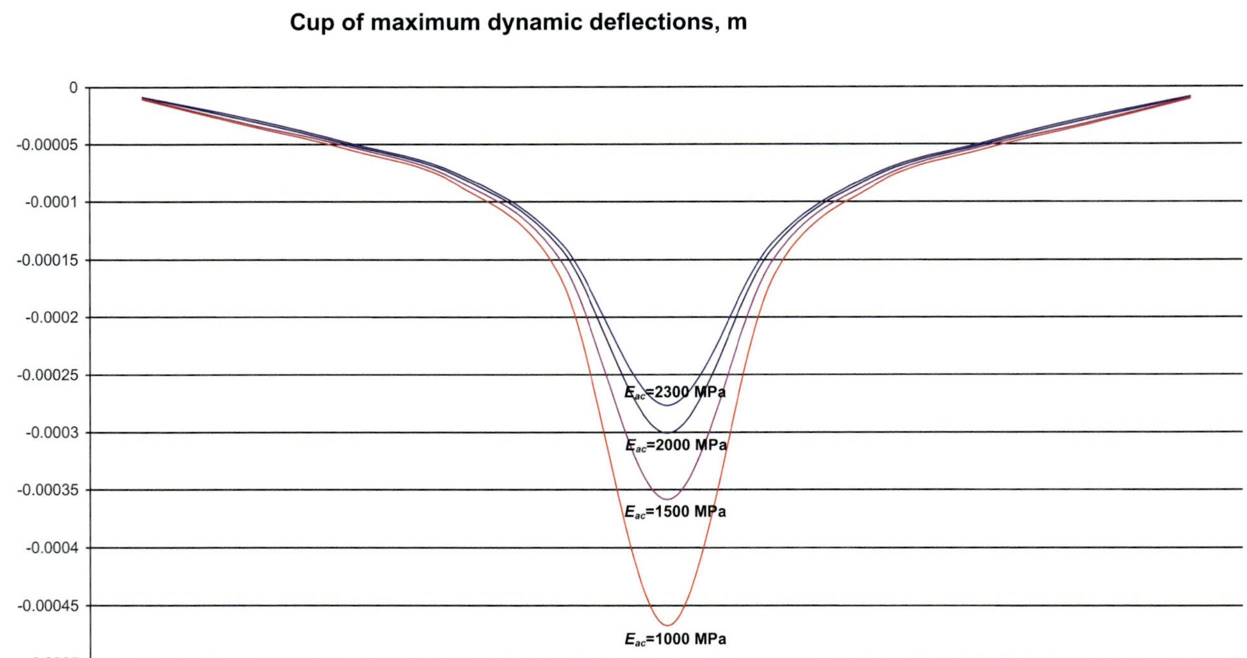

	2.5	1.25	0.75	0.5	0.25	0	0.25	0.5	0.75	1.25	2.5
Construction No.1	-0.000008504	-0.000027156	-0.000049513	-0.000075563	-0.000136273	-0.000276686	-0.000136273	-0.000075563	-0.000049513	-0.000027156	-0.000008504
Construction No.1(1)	-0.000008708	-0.000028956	-0.000050859	-0.000077592	-0.000141277	-0.000300621	-0.000141277	-0.000077592	-0.000050859	-0.000028956	-0.000008708
Construction No.1(2)	-0.000009535	-0.000032006	-0.000052585	-0.000081305	-0.000152214	-0.000358537	-0.000152214	-0.000081305	-0.000052585	-0.000032006	-0.000009535
Construction No.1(3)	-0.000010621	-0.000033605	-0.000055243	-0.000085054	-0.000168329	-0.000466691	-0.000168329	-0.000085054	-0.000055243	-0.000033605	-0.000010621

Fig. 10.14 Simulation of the calculated characteristics of coating

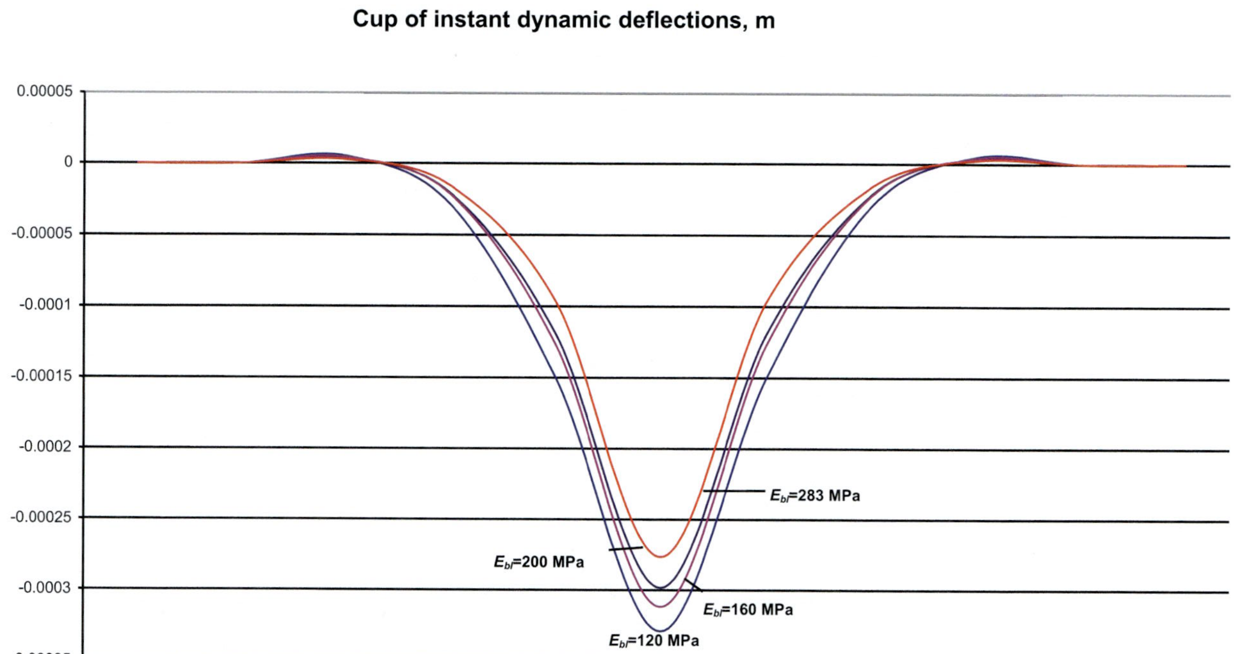

	2.5	1.25	0.75	0.5	0.25	0	0.25	0.5	0.75	1.25	2.5
Construction No.3	0.0000000033	-0.000000091	0.0000049306	-0.000029660	-0.000150796	-0.000328594	-0.000150796	-0.000029660	0.0000049306	-0.000000091	0.0000000033
Construction No.3(1)	0.0000000047	-0.000000063	0.0000036032	-0.000021585	-0.000120079	-0.000298250	-0.000120079	-0.000021585	0.0000036032	-0.000000063	0.0000000047
Construction No.3(2)	0.0000000082	-0.000000082	0.0000041803	-0.000022522	-0.000127627	-0.000311669	-0.000127627	-0.000022522	0.0000041803	-0.000000082	0.0000000082
Construction No.1	0.0000000040	-0.000000042	0.0000027039	-0.000016242	-0.000097211	-0.000276686	-0.000097211	-0.000016242	0.0000027039	-0.000000042	0.0000000040

Fig. 10.15 Simulation of the calculated characteristics of base

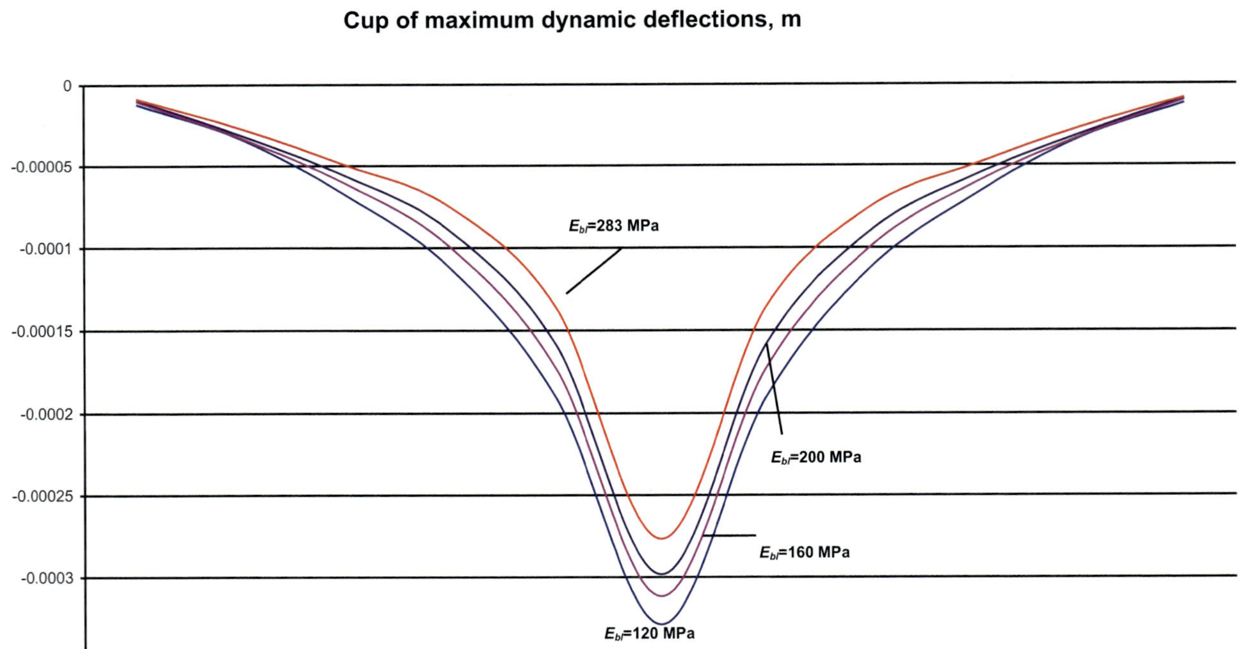

	2.5	1.25	0.75	0.5	0.25	0	0.25	0.5	0.75	1.25	2.5
Construction No.3	-0.000011986	-0.000033407	-0.000068041	-0.000112539	-0.000190843	-0.000328594	-0.000190843	-0.000112539	-0.000068041	-0.000033407	-0.000011986
Construction No.3(1)	-0.000009220	-0.000031186	-0.000056381	-0.000090052	-0.000158802	-0.000298250	-0.000158802	-0.000090052	-0.000056381	-0.000031186	-0.000009220
Construction No.3(2)	-0.000010126	-0.000032732	-0.000061167	-0.000099759	-0.000173049	-0.000311669	-0.000173049	-0.000099759	-0.000061167	-0.000032732	-0.000010126
Construction No.1	-0.000008504	-0.000027156	-0.000049513	-0.000075563	-0.000136273	-0.000276686	-0.000136273	-0.000075563	-0.000049513	-0.000027156	-0.000008504

Fig. 10.16 Simulation of the calculated characteristics of base

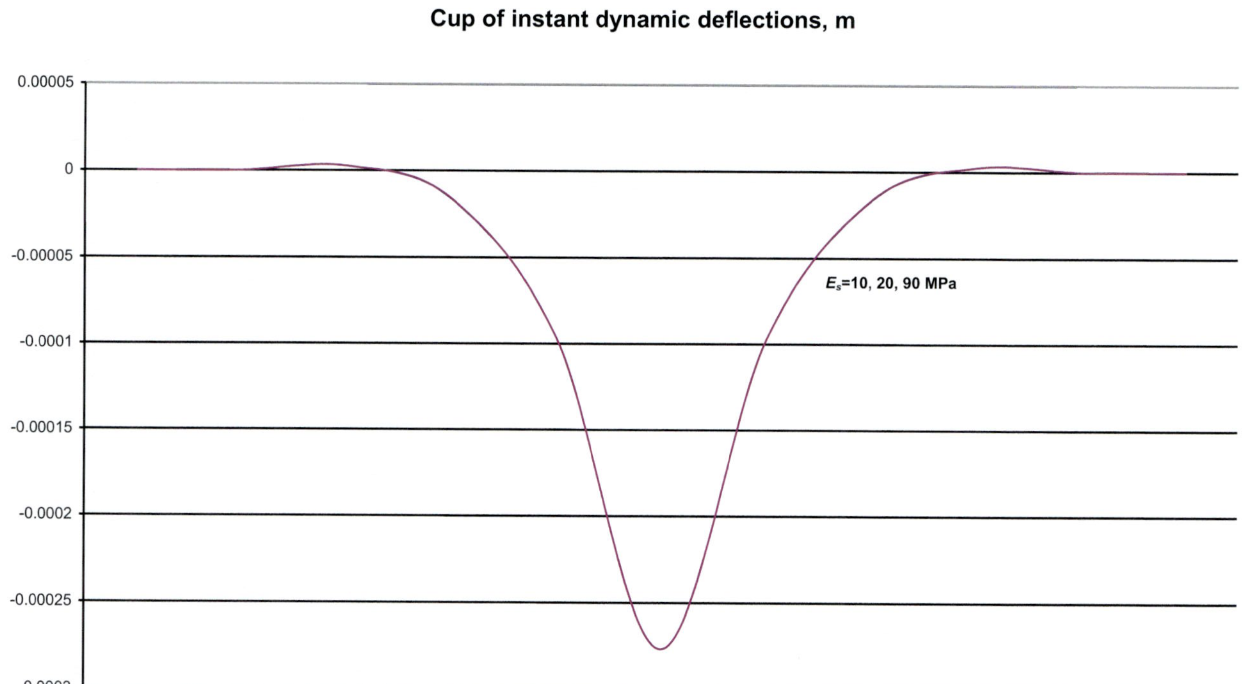

	2.5	1.25	0.75	0.5	0.25	0	0.25	0.5	0.75	1.25	2.5
Construction No.2	0.0000000040	-0.000000042	0.0000027083	-0.000016230	-0.000097214	-0.000276708	-0.000097214	-0.000016230	0.0000027083	-0.000000042	0.0000000040
Construction No.2(1)	0.0000000040	-0.000000042	0.0000026992	-0.000016254	-0.000097212	-0.000276674	-0.000097212	-0.000016254	0.0000026992	-0.000000042	0.0000000040
Construction No.2(2)	0.0000000040	-0.000000042	0.0000027104	-0.000016225	-0.000097219	-0.000276725	-0.000097219	-0.000016225	0.0000027104	-0.000000042	0.0000000040

Fig. 10.17 Simulation of the calculated characteristics of subgrade

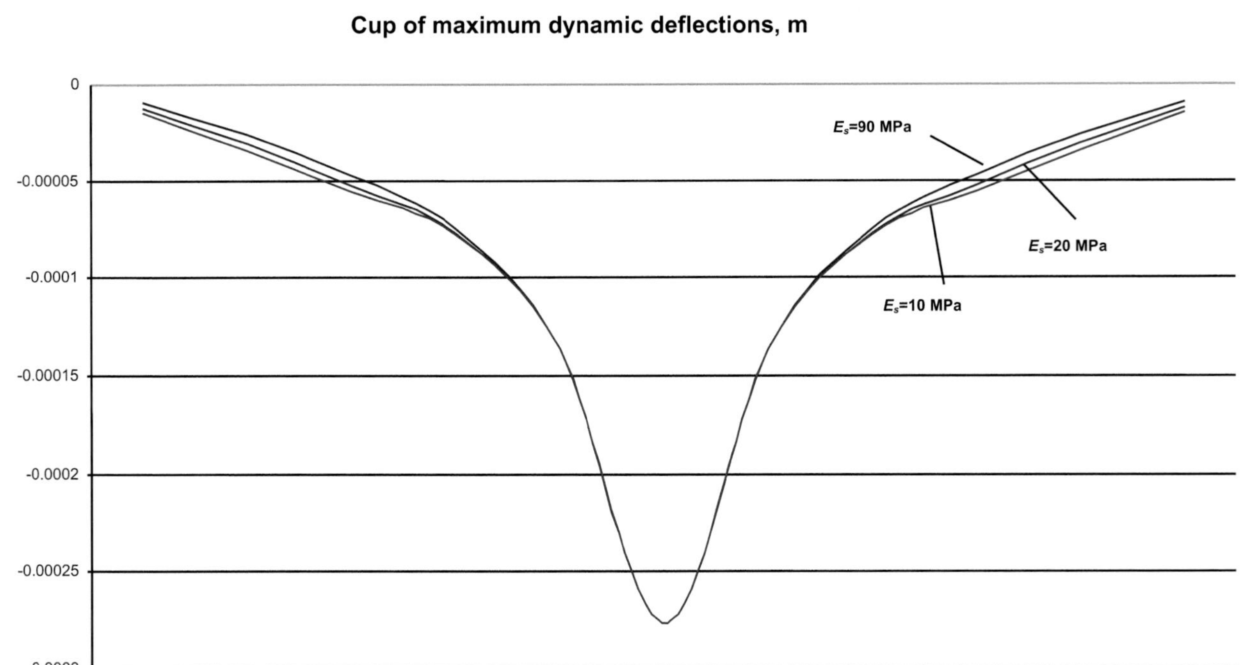

	2.5	1.25	0.75	0.5	0.25	0	0.25	0.5	0.75	1.25	2.5
Construction No.2	-0.000011868	-0.000030751	-0.000052848	-0.000077062	-0.000136096	-0.000276708	-0.000136096	-0.000077062	-0.000052848	-0.000030751	-0.000011868
Construction No.2(1)	-0.000009160	-0.000025929	-0.000046842	-0.000074266	-0.000136432	-0.000276674	-0.000136432	-0.000074266	-0.000046842	-0.000025929	-0.000009160
Construction No.2(2)	-0.000014162	-0.000034084	-0.000055251	-0.000078000	-0.000135997	-0.000276725	-0.000135997	-0.000078000	-0.000055251	-0.000034084	-0.000014162

Fig. 10.18 Simulation of the calculated characteristics of subgrade

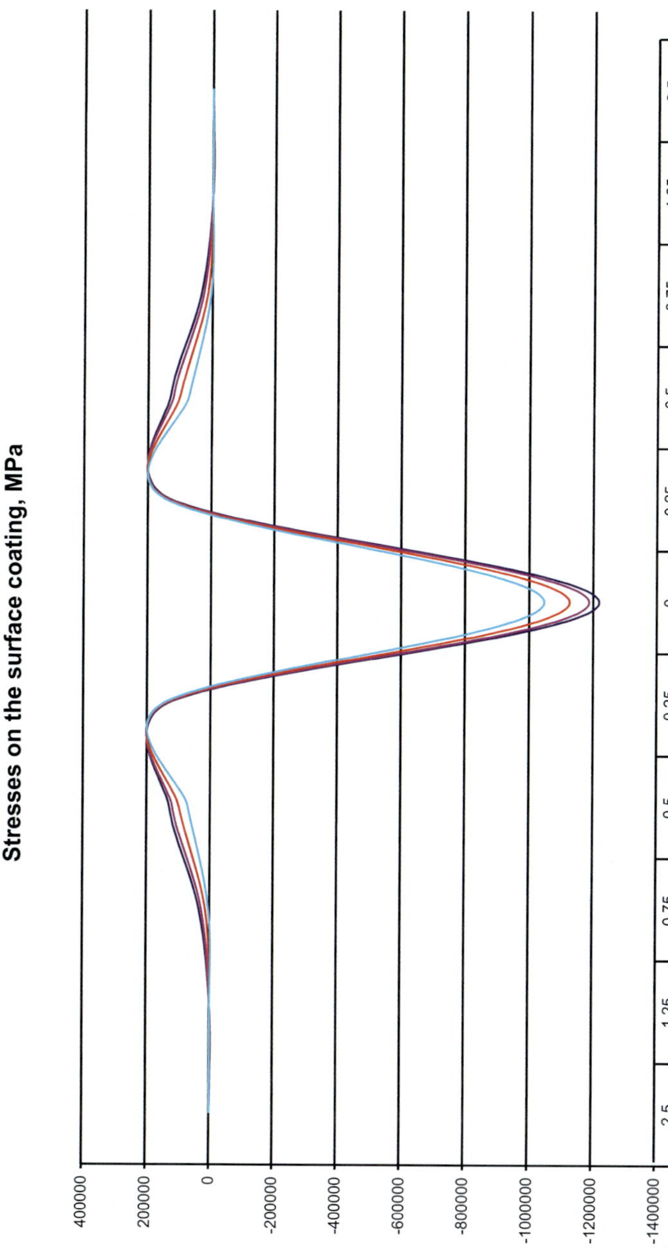

	2.5	1.25	0.75	0.5	0.25	0	0.25	0.5	0.75	1.25	2.5
Construction No.1	-38.346	-1122.8	34856	129843	133801	-1218600	133801	129843	34856	-1122.8	-38.346
Construction No.1(1)	-29.225	-726.5	27317	120027	138318	-1187710	138318	120027	27317	-726.5	-29.225
Construction No.1(2)	-14.359	-323.66	14872	99620	144760	-1126590	144760	99620	14872	-323.66	-14.359
Construction No.1(3)	-3.5182	-146.47	4034.3	71762	148353	-1047360	148353	71762	4034.3	-146.47	-3.5182

Fig. 10.19 Simulation of the calculated characteristics of coating

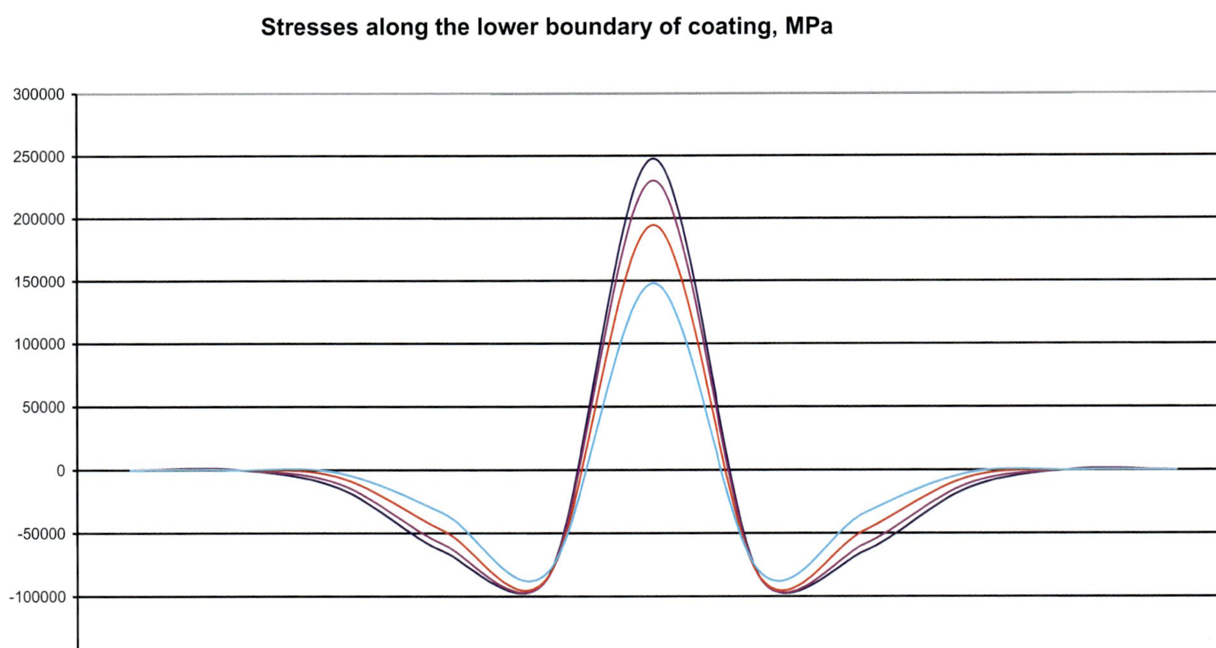

	2.5	1.25	0.75	0.5	0.25	0	0.25	0.5	0.75	1.25	2.5
Construction No.1	-21.181	398.09	-14398	-64617	-83104	247455	-83104	-64617	-14398	398.09	-21.181
Construction No.1(1)	-18.027	164.24	-11093	-59247	-83672	229809	-83672	-59247	-11093	164.24	-18.027
Construction No.1(2)	-9.1252	-65.643	-6004.4	-48419	-83222	194460	-83222	-48419	-6004.4	-65.643	-9.1252
Construction No.1(3)	-1.9135	-78.722	-2333.6	-34587	-79497	147907	-79497	-34587	-2333.6	-78.722	-1.9135

Fig. 10.20 Simulation of the calculated characteristics of coating

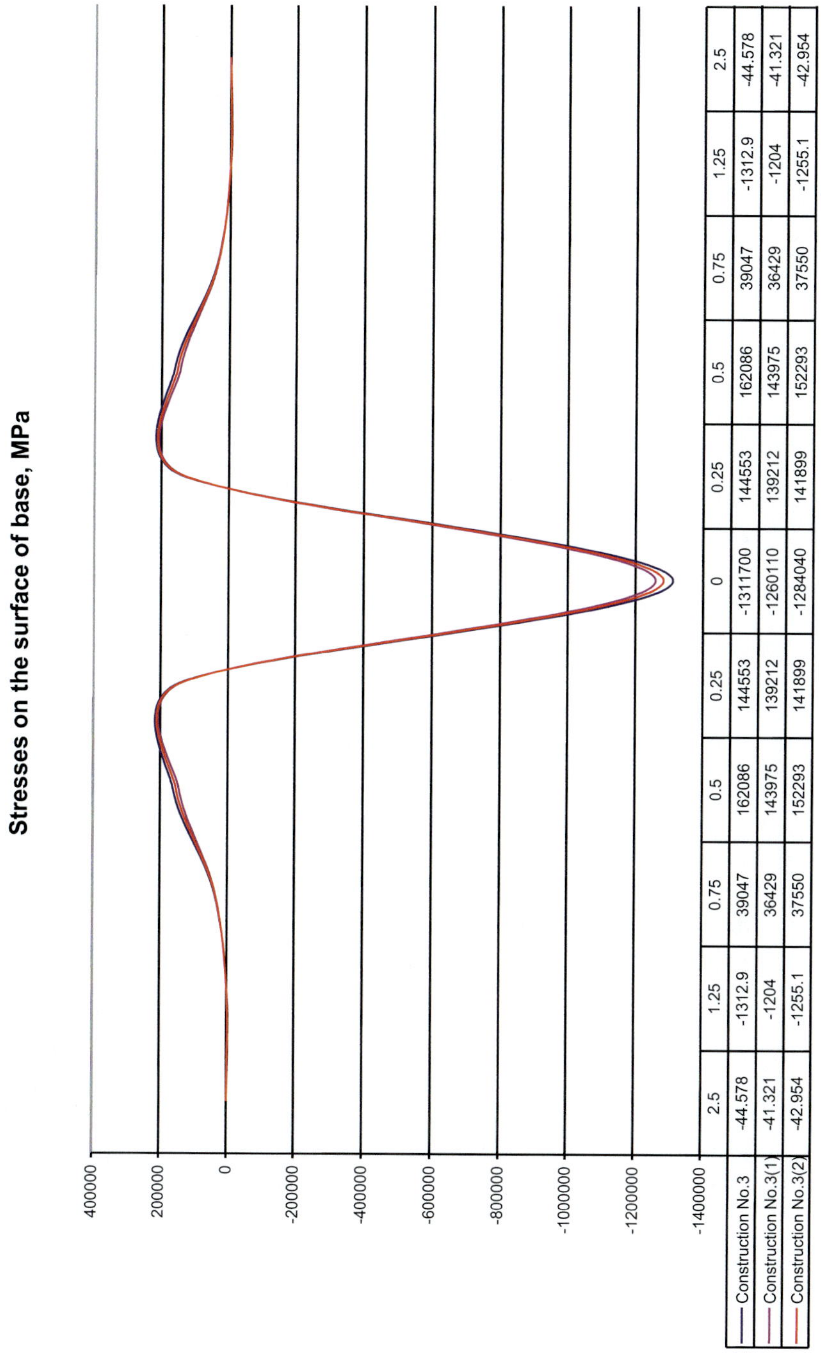

	2.5	1.25	0.75	0.5	0.25	0	0.25	0.5	0.75	1.25	2.5
Construction No.3	-44.578	-1312.9	39047	162086	144553	-1311700	144553	162086	39047	-1312.9	-44.578
Construction No.3(1)	-41.321	-1204	36429	143975	139212	-1260110	139212	143975	36429	-1204	-41.321
Construction No.3(2)	-42.954	-1255.1	37550	152293	141899	-1284040	141899	152293	37550	-1255.1	-42.954

Fig. 10.21 Simulation of the calculated characteristics on interface between coating and base

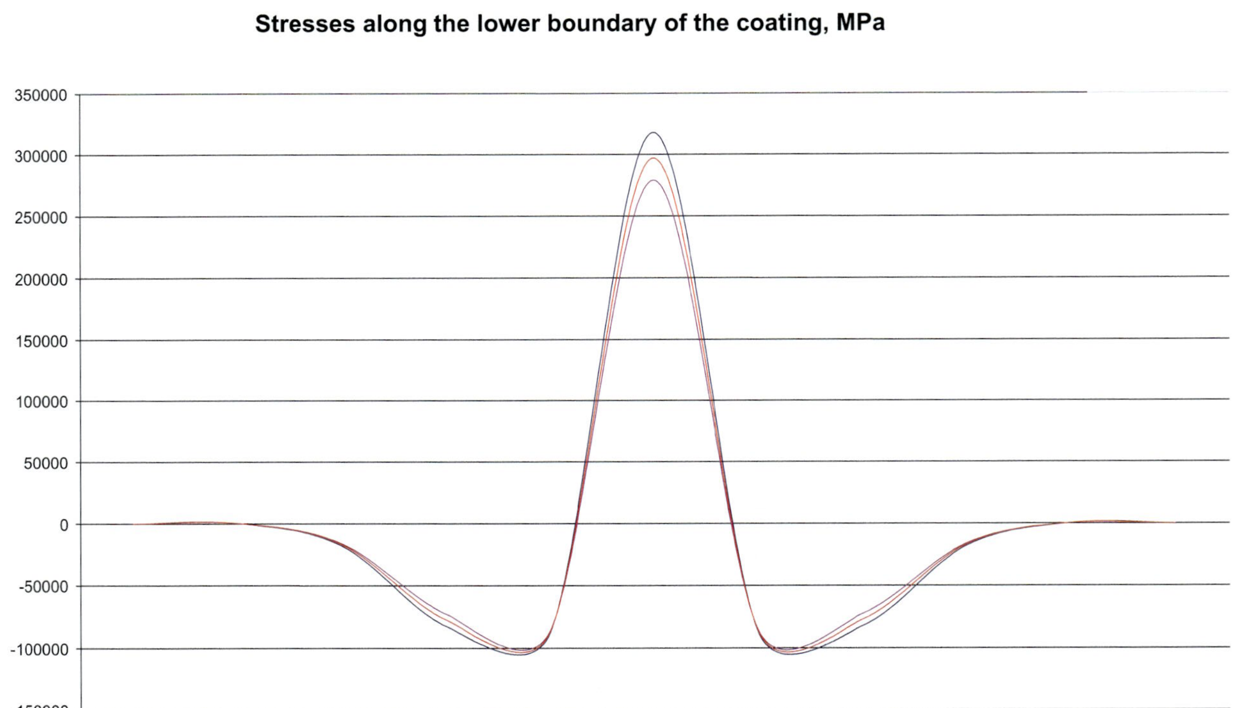

	2.5	1.25	0.75	0.5	0.25	0	0.25	0.5	0.75	1.25	2.5
Construction No.3	-25.583	352.25	-17534	-82200	-87639	318042	-87639	-82200	-17534	352.25	-25.583
Construction No.3(1)	-23.562	375.76	-15690	-72456	-85661	279007	-85661	-72456	-15690	375.76	-23.562
Construction No.3(2)	-24.634	363.55	-16515	-76985	-86774	297144	-86774	-76985	-16515	363.55	-24.634

Fig. 10.22 Simulation of the calculated characteristics on interface between coating and base

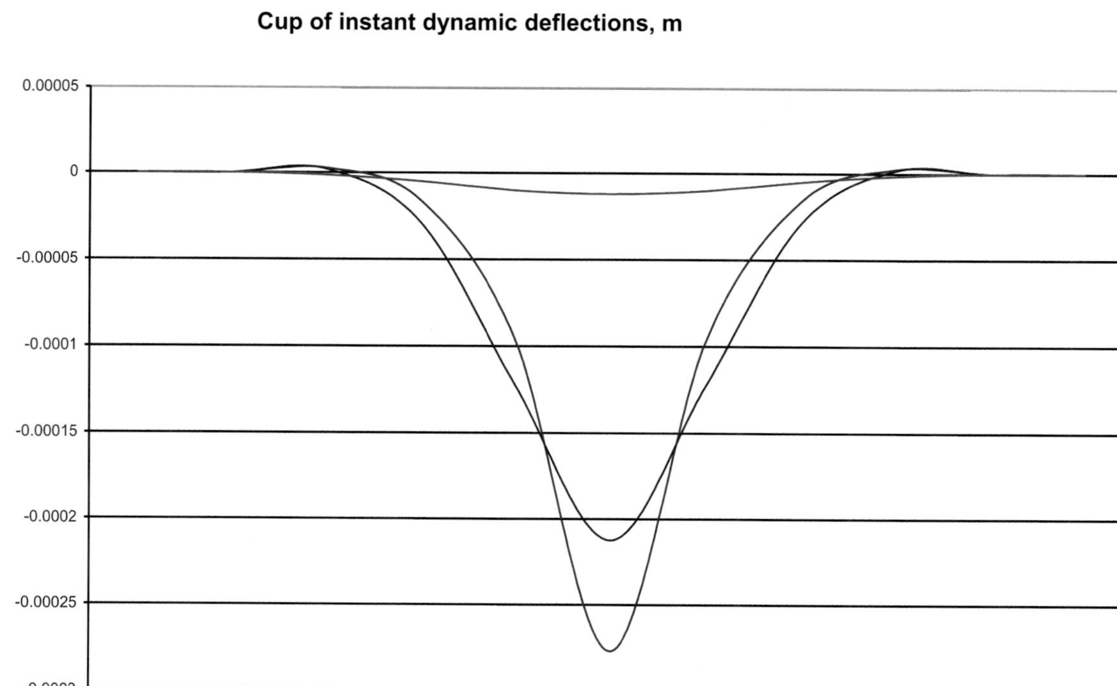

	2.5	1.25	0.75	0.5	0.25	0	0.25	0.5	0.75	1.25	2.5
Construction No.1(on surface)	0.000000004	-0.00000004	0.000002703	-0.00001624	-0.00009721	-0.00027668	-0.00009721	-0.00001624	0.000002703	-0.00000004	0.000000004
Construction No.1(under coating)	-0.00000002	0.000000067	0.000001946	-0.00002911	-0.00012288	-0.00021251	-0.00012288	-0.00002911	0.000001946	0.000000067	-0.00000002
Construction No.1(under base)	-0.00000000	0.000000007	-0.00000134	-0.00000502	-0.00000996	-0.00001184	-0.00000996	-0.00000502	-0.00000134	0.000000007	-0.00000000

Fig. 10.23 Change of instant cup of dynamic deflection along horizontal sections in construction No. 1

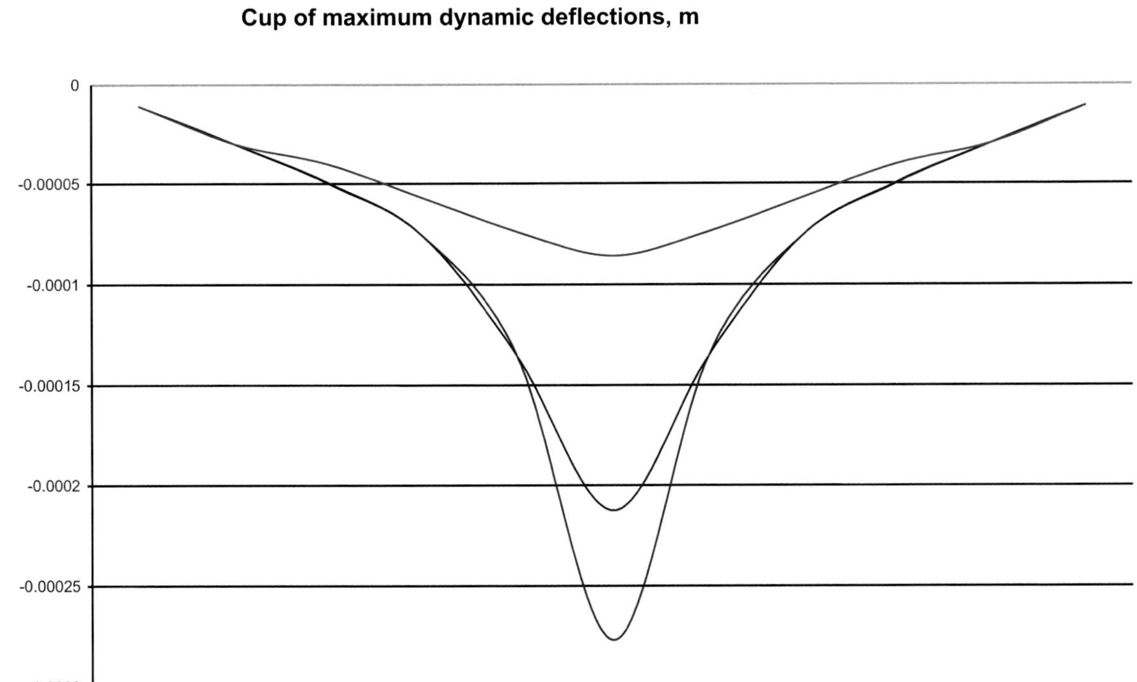

	2.5	1.25	0.75	0.5	0.25	0	0.25	0.5	0.75	1.25	2.5
Construction No.1(on surface)	-0.000011	-0.000030	-0.000050	-0.000076	-0.000136	-0.000277	-0.000136	-0.000076	-0.000050	-0.000030	-0.000011
Construction No.1(under coating)	-0.000011	-0.000030	-0.000050	-0.000076	-0.000136	-0.000213	-0.000136	-0.000076	-0.000050	-0.000030	-0.000011
Construction No.1(under base)	-0.000011	-0.000030	-0.000040	-0.000057	-0.000074	-0.000086	-0.000074	-0.000057	-0.000040	-0.000030	-0.000011

Fig. 10.24 Change of maximum cup of dynamic deflection along horizontal sections in construction No. 1

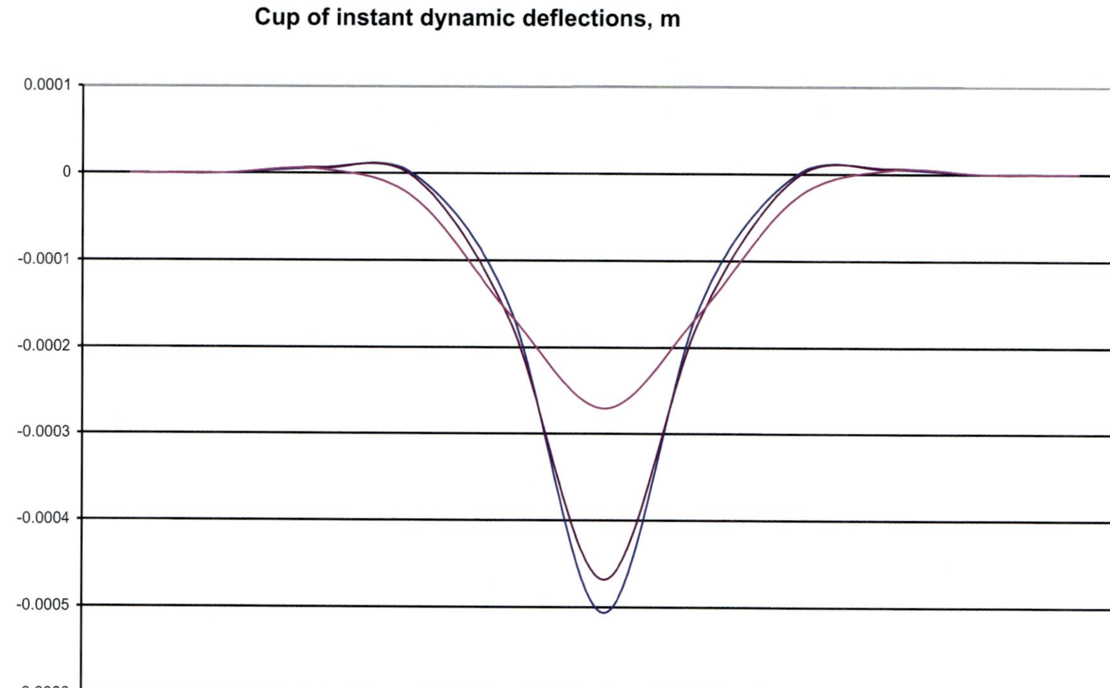

	2.5	1.25	0.75	0.5	0.25	0	0.25	0.5	0.75	1.25	2.5
Construction No.4(on surface)	0.000000004	-0.00000001	0.000005447	-0.00000235	-0.00015289	-0.00050614	-0.00015289	-0.00000235	0.000005447	-0.00000001	0.000000004
Construction No.4(under coating)	0.000000009	-0.00000003	0.000006646	-0.00000530	-0.00016788	-0.00046821	-0.00016788	-0.00000530	0.000006646	-0.00000003	0.000000009
Construction No.4(under base)	0.000000003	-0.00000003	0.000004527	-0.00002855	-0.00016119	-0.00027068	-0.00016119	-0.00002855	0.000004527	-0.00000003	0.000000003

Fig. 10.25 Change of instant cup of dynamic deflection along horizontal sections in construction No. 4

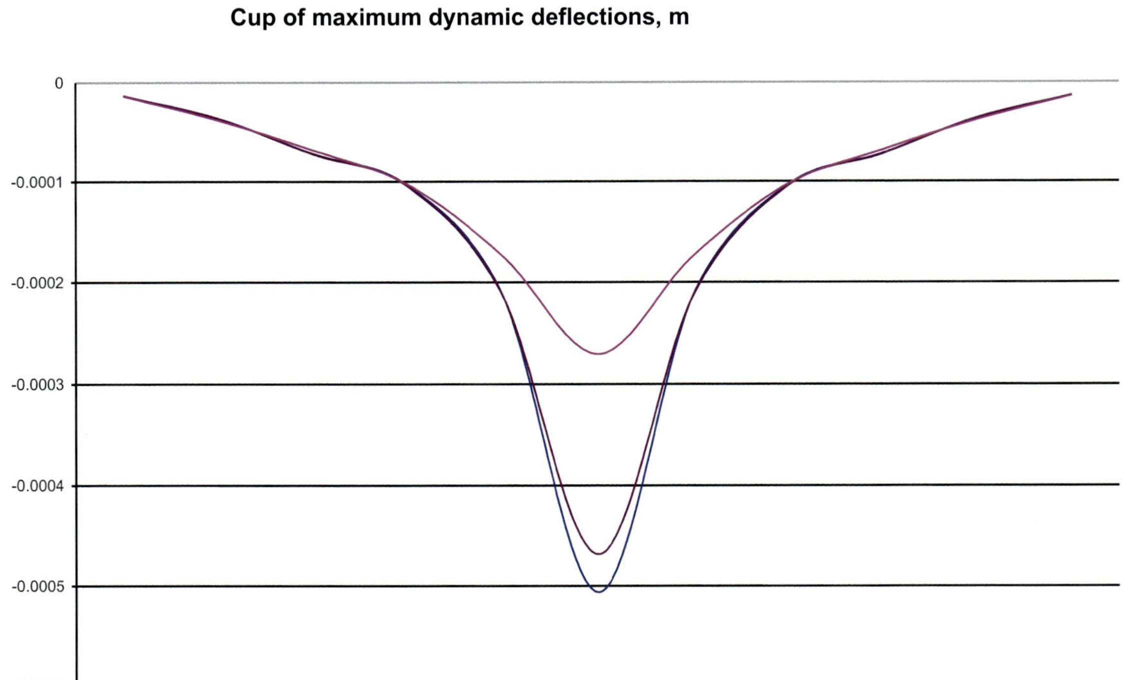

	2.5	1.25	0.75	0.5	0.25	0	0.25	0.5	0.75	1.25	2.5
Construction No.4(on surface)	-0.00001318	-0.00003594	-0.00007190	-0.00010383	-0.00021327	-0.00050614	-0.00021327	-0.00010383	-0.00007190	-0.00003594	-0.00001318
Construction No.4(under coating)	-0.00001316	-0.00003608	-0.00007238	-0.00010466	-0.00021375	-0.00046821	-0.00021375	-0.00010466	-0.00007238	-0.00003608	-0.00001316
Construction No.4(under base)	-0.00001365	-0.00003791	-0.00006829	-0.00010279	-0.00017410	-0.00027068	-0.00017410	-0.00010279	-0.00006829	-0.00003791	-0.00001365

Fig. 10.26 Change of maximum cup of dynamic deflection along horizontal sections in construction No. 4

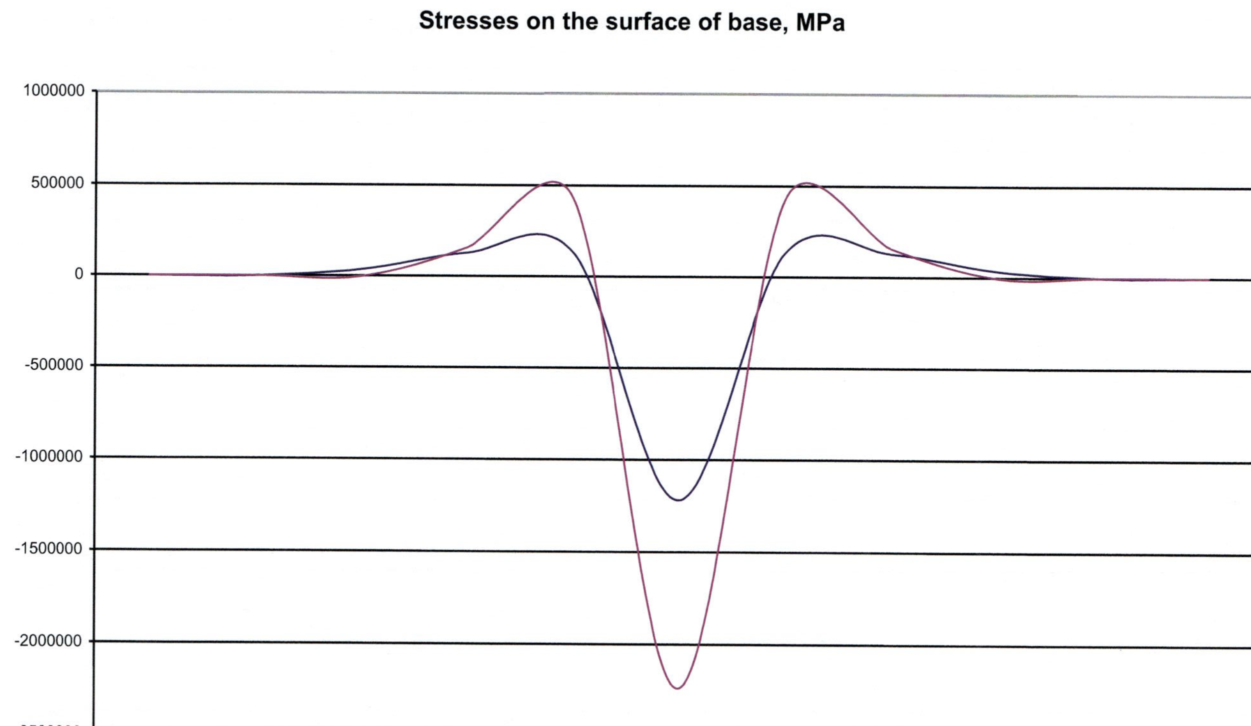

	2.5	1.25	0.75	0.5	0.25	0	0.25	0.5	0.75	1.25	2.5
Construction No.1	-38.346	-1122.8	34856	129843	133801	-1218600	133801	129843	34856	-1122.8	-38.346
Construction No.4	1.283	463.83	-3055.4	158912	429883	-2238920	429883	158912	-3055.4	463.83	1.283

Fig. 10.27 Stresses on the interface between coating and base

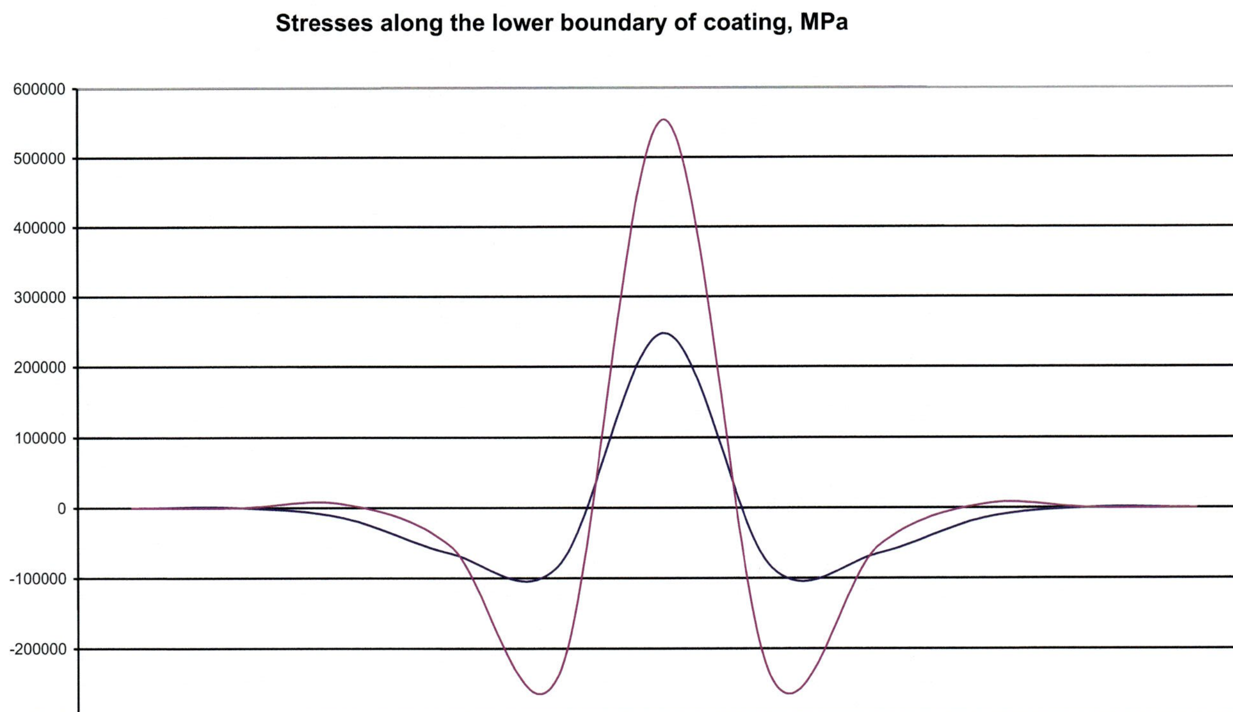

	2.5	1.25	0.75	0.5	0.25	0	0.25	0.5	0.75	1.25	2.5
Construction No.1	-21.181	398.09	-14398	-64617	-83104	247455	-83104	-64617	-14398	398.09	-21.181
Construction No.4	2.0034	-84.093	5450.4	-55017	-240520	554564	-240520	-55017	5450.4	-84.093	2.0034

Fig. 10.28 Stresses on the interface between coating and base

tion elements, we considered namely the plots of the cups of the maximum deflections along the horizontal cross sections of the construction. Based on the analysis of these plots, the contribution of deformation of structural elements to the complete deformation of the surface of the structure was revealed. The stress distribution in road constructions, including tensile forces near the upper and lower boundaries of the asphalt concrete layer, was also calculated. The most characteristic plots of the stress distributions in constructions (including tensile forces) are shown in Figs. 10.27 and 10.28.

Analysis and systematization of the results of the numerical experiments allow us to state the following:

1. The road construction of enhanced strength (with a thickness of asphalt concrete layers of more than 18 cm) is characterized by significantly lower displacements in the zone at the distance of up to 0.75 m. At distances of 1.25–2.5 m, displacements for different structures are close, because subgrade soil has the same modulus of elasticity.
2. For a weaker construction of pavement (with a thickness of asphalt concrete layers equal to 10 cm), a rapid increase in the value of the parameters in the zone close to the impact (at a distance of 0.25 m) is characteristic, which indicates a low strength of the asphalt concrete pavement.
3. The study of dynamic deflections along the boundaries of the elements of the road construction (pavement, base, soil of the subgrade) shows that in this construction, the deformation leading to strengthening of the subgrade soil is 35% of the total deformation; 40% of the strain appears due to compression of the base layers and 25% is caused by the compression of asphalt concrete layers (Fig. 10.24). For a weaker construction, the strain ratio is as follows: the strain in the subgrade soil is 54% of the total strain; 40% of deformation appears due to compression of the base layers and only 6% is caused by the compression of the coating layers (Fig. 10.25).
4. It should be noted that there is a significant delay in the development of in-depth deformation characteristic of a strength construction (on an instant cup of deflection, the value of deformation of the soil of the subgrade is eight times lower than on the maximum cup of deflection). For weak constructions, almost instantly spreading deformation in depth is characteristic.
5. Tensile stresses calculated along the lower boundary of the coating and near the coating surface have higher values for a weak construction (1.5 and 3.0 times, respectively).
6. An increase in the elastic modulus of asphalt concrete leads to an increase in tensile stresses for a strength construction and to an increase in tensile stresses at the lower and upper boundaries for weak constructions (Figs. 10.19 and 10.20).
7. An increase in the elastic modulus of the base leads to a decrease in tensile stresses (Figs. 10.21 and 10.22).

10.4 Concluding Remarks

1. Developed and implemented mechanical and mathematical models of the system of "road construction-soil" under test dynamic impact, simulating a hard cup on the surface of the coating with various levels of complexity.
2. The fields of application of the multilayer construction models developed by using FEM are identified. The correspondence of the results of calculating the wave and energy characteristics of stationary and nonstationary analysis at plane and spatial deformations of the medium with the results obtained by the analytical method is shown.
3. It has been established that the cups of maximum deflections are more depending on the ratios of the elastic moduli of the structural elements of the system. An analysis of the results of a numerical experiment made it possible to establish the following regularities in the formation of a cup of maximum deflections of the surface of a coating under impact loading:

(i) A decrease in the elastic modulus of the asphalt concrete coating leads to significant changes in the characteristics of the cup of maximum deflections in the zone closest to the impact point (up to 0.25 m); it is manifested in an increase in maximum amplitudes and a decrease in the radius of curvature of the surface of the coating in this zone, and as a result leads to an increase in tensile stresses in coating.
(ii) A decrease in the elastic modulus of the base is manifested in a change in the cup of maximum deflections in the zone of 0.25–0.75 m.
(iii) A decrease in the elastic modulus of the soil of the subgrade affects the change in the shape of the oscillation cup in the far zone from the impact point at a distance of more than 0.75 m.

These patterns determine the criterion for the placements of observation points on the surface of the structure that is at least 4 points at a distance of 0.25, 0.75, 1.25, and 2.5 m from the center of impact. In this case, it is possible to accurately restore the shape of the cup of the maximum deflections of the road construction.

References

1. Adeeva, L. N., & Borbat, L. N. (2009). *Bulletin of the Omsk University, 2*, 141. (In Russian).
2. Alekhin, A. A. (1995). *Gorny Journal, 1*, 116.
3. Arioglu, E., Kezhen, L., & Kayyanyan, S. (1987). *Development of deposits with a bookmark*. Moscow: Mir. (In Russian).
4. Babeshko, V. A., Glushkov, E. V., & Zinchenko, Z. F. (1989). *Dynamics of nonhomogeneous linear elastic medium*. Moscow: Nauka. (In Russian).
5. Babkov, V. V., Sakhibgariev, R. R., & Chuykin, N. E. (2013). *Building Materials, 10*, 23. (In Russian).
6. Balabanov, B. B., & Nikolaenko, V. P. (2011). *Bulletin of IrGTU, 6*(53), 37. (In Russian).
7. Bazhenov, Y. M., Alimov, L. A., & Voronin, V. V. (1996). *News of Universities. Construction, 7*, 55. (In Russian).
8. Bazhenov, Y. M., Alimov, L. A., & Voronin, V. V. (1997). *News of Universities. Construction, 4*, 68. (In Russian).
9. Bazhenov, Y. M., Voronin, V. V., & Alimov, L. A. (2013). *The structure and properties of concrete with nanomodifiers based on industrial waste*. Moscow: Moscow State Building University Press. (In Russian).
10. Belyaeva, L. S., & Snisarenko, R. F. (1992). *Ugol Ukrainy, 11*, 48. (In Russian).
11. Bent, O. I. (1992). Mineral Resources of Russia. *Economics and Management, 6*, 26. (In Russian).
12. Bent, O. I. (1994). *Building Materials and Structures, 2*, 28. (In Russian).
13. Boldyrev, V. V., Avvakumov, E. G., & Boldyreva, E. V. (2009). *Fundamentals of mechanical activation, mechanosynthesis and mechanical technology*. Novosibirsk: Publishing House of the SB RAS. (In Russian).
14. Borisenko, L. F., Delitsyn, L. M., & Vlasov, A. S. (2001). *Prospects for the use of ash from coal-fired thermal power plants*. Moscow: Geoinformmark. (In Russian).
15. Bracewell, R. (1986). *The Fourier transform and its applications*. New York: McGraw-Hills.
16. Brailovsky, S. S., & Shain, A. M. (1990). *Highways, 4*, 22. (In Russian).
17. Brekhovskikh, L. M. (1957). *Waves in layered media*. Moscow: USSR Academy of Sciences Press. (In Russian).
18. Buravchuk, N. I. (2009). *Resource conservation in construction materials technology*. SFedU: Rostov-on-Don. (In Russian).
19. Buravchuk, N. I., Budnitsky, V. M., & Brazhnikov, V. F. (1998). *University News. North Caucasus Region. Technical Science, 4*, 70. (In Russian).

A. A. Lyapin et al., *Improving Road Pavement Characteristics*, Innovation and Discovery in Russian Science and Engineering,
https://doi.org/10.1007/978-3-030-59230-1

20. Buravchuk, N. I., Budnitsky, V. M., Brazhnikov, V. F., & Melentiev, S. A. (1999). *Resource conservation in the technology of binders and concrete*. Rostov-on-Don: North Caucasus Scientific Centre of High School Press. (In Russian).
21. Buravchuk, N. I., & Guryanova, O. V. (2014). *Ecological Bulletin of Russia, 8*, 46. (In Russian).
22. Buravchuk, N. I., & Guryanova, O. V. (2015). *News of Universities. North Caucasus Region. Engineering, 1*, 111. (In Russian).
23. Buravchuk, N. I., & Guryanova, O. V. (2016). *News of Universities. North Caucasus Region. Technical. Sciences, 3*, 104. (In Russian).
24. Buravchuk, N. I., & Guryanova, O. V. (2017). *Composition for obtaining a fuel briquette*: Russian Patent No. 2664567, 08.29.2017. (In Russian).
25. Buravchuk, N. I., & Guryanova, O. V. (2018). *Ecological Bulletin of Russia, 5*, 42. (In Russian).
26. Buravchuk, N. I., & Guryanova, O. V. (2018). *Proceedings of the International Science Conference "Advances of Science". Czech Republic: Karlovy Vary. Russia Moscow. March 29–30, 2018.*
27. Buravchuk, N. I., & Guryanova, O. V. (2019). *Innovation and Expertise, 2*(23), 195. (In Russian).
28. Buravchuk, N. I., Guryanova, O. V., & Hayrapetyan, M. A. (2007). *News of Universities. North Caucasus Region. Technical. Sciences, 4*, 100. (In Russian).
29. Buravchuk, N. I., Guryanova, O.V., Jani, M.A., & Putri, E. P. (2018). In: *Advanced Materials - Proceedings of the International Conference on "Physics and Mechanics of New Materials and Their Applications", PHENMA 2017,* Springer Proceedings in Physics, Ivan A. Parinov, Shun–Hsyung Chang, Vijay K. Gupta (Eds.). Switzerland: Springer Nature. Cham, *207*, 640.
30. Buravchuk, N. I., Guryanova, O.V., Jani, M.A., & Putri, E.P. (2019). In: *Advanced Materials - Proceedings of the International Conference on "Physics and Mechanics of New Materials and Their Applications", PHENMA 2018,* Springer Proceedings in Physics, Ivan A. Parinov, Shun–Hsyung Chang. Yun-Hae Kim (Eds.). Switzerland: Springer Nature. Cham, *224*, 666.
31. Buravchuk, N. I., Guryanova, O. V., Okorokov, E. P., & Pavlova, L. N. (2001). In: *Proceedings of the X All-Russian Coal Conference.* Rostov-on-Don: VNIGRIugol, (In Russian).
32. Buravchuk, N. I., Guryanova, O. V., Okorokov, E. P., & Pavlova, L. N. (2010). *Ecological Bulletin of Russia, 5*, 26. (In Russian).
33. Buravchuk, N. I., Guryanova, O. V., Okorokov, E. P., & Pavlova, L. N. (2011). *Ecological Bulletin of Russia, 11*, 44. (In Russian).
34. Buravchuk, N. I., Guryanova, O. V., & Pack, G. N. (2014). *News of Universities. North Caucasus Region. Engineering, 1*, 75. (In Russian).
35. Buravchuk, N. I., & Kieselstein, L.Ya. (2002). In: *Problems of the integrated use of coal deposits. Proceedings of the All-Russian Scientific and Technical Seminar*. Rostov-on-Don. VNIGRIugol, (In Russian).
36. Buravchuk, N. I., Kondyurin, A. M., & Guryanova, O. V. (2012). *Bulletin of V.G. Shukhov Belgorod State Technical University, 4*, 11. (In Russian).
37. Buravchuk, N. I., Myshev, M. P., & Guryanova, O. V. (1991). *New methods for the comprehensive study and use of coal, oil shale and related minerals for exploration. Collection of scientific papers of VIMS*. Moscow. (In Russian).
38. Buravchuk, N. I., & Rutkov, K. I. (1997). *Processing and use of solid mining waste and burning waste*. Rostov-on-Don: North Caucasus Scientific Centre of High School Press. (In Russian).
39. Butkevich, G. R. (1996). *Building Materials, 4*, 8. (In Russian).
40. Ceccherini-Silberstein, T., Scarabotti, F., & Tolli, F. (2018). *Discrete harmonic analysis*. Cambridge: Cambridge University Press.
41. Cherepanov, A. A., & Kardash, V. T. (2009). *Geology and Minerals of the World Ocean, 2*, 98. (In Russian).
42. Chernyshev, E. M., & Dyachenko, E. I. (1996). *University News. Building., 3*, 43. (In Russian).
43. Chesnokov, B. V., & Shcherbakova, E. P. (1991). *Mineralogy of the burnt dumps of the Chelyabinsk coal basin. The mineralogy of technogenesis experience*. Moscow: Nauka. (In Russian).

44. Chumachenko, N. G. (2011). *Bulletin of SASAC. Urban Planning and Architecture, 1*, 112. (In Russian).
45. Chumachenko, N. G., Turnikov, P. V., Seikin, A. I., & Bannova, S. E. (2015). *Ecology and Industry of Russia, 19*(11), 41. (In Russian).
46. Deryugin, B. V., Nerpin, C. B., & Churaev, N.V. (1974). In: *Proceedings of the X International. Congress of Soil Scientists,* Moscow: Science, *I*. (In Russian).
47. Deryugin, B.V., Zhelezny, B.V., Zorin, Z.M., Sobolev, V.D., Churaev. N.V. (1974) In: *Proceedings of the 5th Conference on Surface Forces.* Moscow: Science. (In Russian).
48. Dolgopolov, V. M., Alekhin, A. A., & Bogdanov, N. M. (1990). *Construction Materials, 12*, 10. (In Russian).
49. Duz, A. I., Pichugin, B. V., & Dudenko, I. I. (1990). *Environmental protection and use of coal production waste*. Donetsk: Donbass. (In Russian).
50. Dvorkin, L. I., & Dvorkin, O. L. (2007). *Building Materials from Industrial Wastes: A Training Manual*. Phoenix: Rostov-on-Don. (In Russian).
51. (1997). "Flat as a Pancake" pavement meets tough spees in Kansas. ENR: Engineering News-Record, *18*, 238.
52. Ermolaev, V. A., Bedrina, G. P., Zervandova, V. P., & Moseikin, V. V. (1996). *News of Higher Education Institutions. Geology and Intelligence, 6*, 65. (In Russian).
53. Euler, R. (1979). *The chemistry of silica*. New York: Wiley.
54. Fedorenko, A. I., Eismot, S. N., & Ushakov, B. C. (1990). *Complex Use of Mineral Raw Materials, 4*, 34. (In Russian).
55. Fedorova, N. V., & Rogatina, Y. N. (2005). *News of Universities. North Caucasus Region. Technical. Sciences, 4*(132), 53. (In Russian).
56. Fedorova, N. V., & Shaforost, D. A. (2014). *Ecology of Industrial Production, 3*, 15. (In Russian).
57. Fedorova, N. V., & Shaforost, D. A. (2015). *Teploenergetika, 1*, 53. (In Russian).
58. Fish, I. A. (2003). *News of Universities. Construction, 3*, 45. (In Russian).
59. Fomina, E. Y., & Artemova, O. S. (2011). *Mining Information and Analytical Bulletin, 8*, 273. (In Russian).
60. Forester, V. S. (1997). *Problems of building materials science and new technologies / Collection of reports of the International Conference*. Belgorod: BelGTASM, Part 5.
61. Frolov, Y. K. (2007). *Mining Information and Analytical Bulletin, 7*, 24. (In Russian).
62. Galibina, E. A., & Kreerman, T. B. (1966). *Building Materials*, 5. (In Russian).
63. Gamalius, E. A. (2008). *Bulletin of SUSU. Construction and Architecture, 7*, 22. (In Russian).
64. Gamaliy, E. A., Gorbunov, S. P., & Trofimov, B. Y. (2007–2008). *Progressive materials and technologies in modern construction: International collection of scientific papers*. Novosibirsk: SEA. (In Russian).
65. Gedeonov, P. P., & Yudin, L. V. (1994). *Building Materials, 2*, 16. (In Russian).
66. Grinchenko, V. T., & Meleshko, V. V. (1981). *Harmonic oscillations and waves in elastic bodies*. Kiev: Naukova Dumka. (In Russian).
67. (1986). *Guidelines for the economic assessment of damage to the national economy caused by the negative impact of coal enterprises on the geological environment*. Perm: VNIIOSugol. (In Russian).
68. (2019). *Guidelines for the use of fly ash and ash and slag mixtures from coal combustion at thermal power plants in road construction: ODM 218.2.031–2013*. Omsk: Siberian State Automobile and Highway Academy. (In Russian).
69. Gupta, S., Hussein, M. F. M., Degrande, G., Hunt, H. E. M., & Clouteau, D. (2007). *Soil Dynamic. Earthquake Engineering., 27*(7), 608.
70. Gusev, B. V., Malinina, L. A., & Shcheblykina, T. P. (1997). *Concrete and Reinforced Concrete, 5*, 5. (In Russian).
71. Hypich, L.V. (1998). *Features of the material composition of dump rocks in mines of the East Donbass and new directions for their use*. PhD Thesis, Rostov-on-Don. (In Russian).

72. Hypich. L.V. (2001). In: *Proceedings of the X All-Russian Coal Conference*. Rostov-on-Don: VNIGRIugol. (In Russian).
73. Kadomtsev, M. I., & Lyapin, A. A. (2018). *IOP Conference Series: Material Science and Engineering, 456*, 012070.
74. Kadomtsev, M. I., Lyapin, A. A., & Seleznev, M. G. (2010). *Structural Mechanics and Analysis Constructions, 3*, 61.
75. Kizilstein, L. Y., Dubov, I. V., Spitsgluz, A. L., & Parade, S. G. (1995). *Components of ashes and slags of thermal power plants*. Moscow: Energoatomizdat. (In Russian).
76. Knigina, G. I. (1962). *News of Higher Educational Institutions. Construction and Architecture, 2*, 82. (In Russian).
77. Knigina, G. I. (1966). *Construction materials from burner rocks*. Moscow: Stroyizdat. (In Russian).
78. Knigina, G. I., & Zavadsky, V. F. (1987). *Microcalorimetry of mineral raw materials in the production of building materials*. Moscow: Stroyizdat. (In Russian).
79. Kolbanovskaya, A. S. (1962). *Reports of the USSR Academy of Sciences, 143*(5), 1159. (In Russian).
80. Kornilov, A. V., & Lygina, T. Z. (2005). *Building Materials, 12*, 78. (In Russian).
81. Kosarev, A. I. (1996). *Building Materials, 4*, 12. (In Russian).
82. Kosenko, N. F., & Makarov, V. V. (2008). *Ecology and Industry of Russia, 4*, 44. (In Russian).
83. Kovalenko, L. I., & Omelchenko, N. P. (2009). *Problems of Ecology, 1–2*, 16. (In Russian).
84. Krasavin, A. P. (1991). *Environmental protection in the coal industry*. Moscow: Nedra. (In Russian).
85. Kuzmina, V. P. (2006). *Appendix to the magazine "Building Materials"*, 5, 7 (In Russian).
86. Kuznetsov, V. D. (1998). *News of Universities. Construction, 10*, 73. (In Russian).
87. Lebedikhin, V.A., Sazonov, V.A., & Grandma, I.M. (1981). In: *Proceedings of the Union State "The Use of Stone Materials, Industrial Wastes and Reinforced Soils in Road Structures"*. Moscow. (In Russian).
88. Lee, F. M. (1961). *Chemistry of cement and concrete*. Moscow: Gosstroyizdat. (In Russian).
89. Leonov, P. A., & Surnachev, B. A. (1970). *Rock dumps of coal mines*. Moscow: Nedra. (In Russian).
90. Lesovik, V. S., & Yevtushenko, E. I. (2002). *News of Universities. Construction, 12*, 40. (In Russian).
91. Lyapin, A. A., Seleznev, M. G., Sobisevich, L. E., & Sobisevich, A. L. (1999). *Mechanical and mathematical models in the problems of active seismology*. Moscow: GNITs PGK, Russian Department of Education. (In Russian).
92. Lysikhina, A. I. (1962). *Road coatings and bases using bitumen and tar*. Moscow: Stroyizdat. (In Russian).
93. Madanbekov, N. Z., & Osmonova, B. Z. (2016). *Innovation Science, 12*, 121. (In Russian).
94. Mathews, J., & Walker, R. L. (1970). *Mathematical method of physics*. Reading, MA: Addison Wesley.
95. (1982). *Methods of testing current waste from coal preparation, rock dumps of coal mines and coal processing plants*. Perm, VNIIOSugol of the Ministry of coal industry of the USSR. (In Russian).
96. Mikheyeva, T. E. (1982). *Study of burnt rock to obtain corrosion-resistant plugging solutions*. PhD thesis. Kharkov. (In Russian).
97. Mikheeva, T. E., Galchenko, P. P., Kaplan, B. Ya., & Miroshnichenko, E. G. (1978). *Proceedings of VNIIOMS*. Kharkov, 261. (In Russian).
98. Mironov, A. A., & Bazuev, V. P. (1992). *Design and construction of highways and bridges of Siberia*. Tomsk: Tomsk Civil Engineering Institute Press. (In Russian).
99. Mironov, A. A., & Evgeniev, I. E. (1986). *Highways and environmental protection*. Tomsk: Tomsk State University Press. (In Russian).
100. Miryuk, S. A., & Akhmetov, I. S. (2002). *Building Materials, 9*, 7. (In Russian).
101. Mnukhin, A. G. (2009). *Coal of Ukraine, 5*, 28. (In Russian).

102. Panova, V. F., Panov, S. A., & Karpachev, A. A. (2016). *News of Universities. Construction, 1*, 31. (In Russian).
103. Panteleev, V. G., Larina, E. A., & Melentiev, V. A. (1985). *The composition and properties of ash and slag TPP: Reference guide*. Moscow: Energoatomizdat. (In Russian).
104. Popov, N. A., & Ivanov, I. A. (1961). Proceedings of the West Siberian Branch of the AU and the USSR Academy of Sciences. *Novosibirsk., 5*, 19. (In Russian).
105. Proshin, A. P., Demyanova, V. S., & Kalashnikov, D. V. (2003). *Ecology and Industry of Russia, 8*, 8. (In Russian).
106. Putilin, E. I. (2003). *The use of fly ash and ash-slag mixtures in the construction of roads. Overview of domestic and foreign experience in the application of waste from the combustion of solid fuels at thermal power plants*. Moscow: SOYUZDORNII. (In Russian).
107. Putilov, V. Y. (2013). *Vestnik MEI, 1*, 16. (In Russian).
108. Rafalowski, M. (2003). *Federal Highway Administration,* Washington, DC: U.S. Department of Transportation.
109. Rebinder, P. A. (1979). *Surface phenomena in dispersed systems. Physical and chemical mechanics. Elect. Proc.* Moscow: Nauka. (In Russian).
110. (1986). *Recommendations for the use of ash, slag and ash-slag mixture of thermal power plants in concrete. NIIZHB.* Moscow: Stroyizdat. (In Russian).
111. Rimkevich, V. S., Pushkin, A. A., & Churushova, O. V. (2015). *Mining informational and analytical bulletin, 6*, 250. (In Russian).
112. *Rio Declaration on Environment and Development. Adopted by the United Nations Conference on Environment and Development, Rio-de-Janeiro, June 3–14, 1992.* https://www.un.org/ru/documents/decl_conv/declarations/riodecl.shtml.
113. Rybiev, I. A., & Sokolov, V. G. (1992). *News of Universities. Construction, 5*, 61. (In Russian).
114. Samodurov, S. I., & Rasstegaeva, G. A. (1992). *Design and Construction of Roads and Bridges in Siberia. Collection of Articles*. Tomsk: Tomsk State University. (In Russian).
115. Saybulatov, S. Z., Kapustin, A. P., & Kalmykova, L. F. (1991). *News of Universities. Construction and Architecture, 3*, 60. (In Russian).
116. Segalova, E. E., & Rebinder, P. A. (1960). *Building Materials, 1*, 21. (In Russian).
117. Sekisov, G. V. (1992). *Gorny Journal, 1*, 17. (In Russian).
118. Semenov, V. A. (1991). *Experience in using local natural materials and industrial waste in road construction*. Moscow: Transport. (In Russian).
119. Sergeyev, E. M., Golodkovskaya, G. A., & Ziangirov, R. S. (1983). *Pedology*. Moscow: Moscow State University Press. (In Russian).
120. Sevostyanov, V. S., Bobyrev, I. N., & Perelgin, M. V. (2001). *Proceeding of the Seventh academic readings of RAASN*. Belgorod: Belgorod State Academy of Building Materials. (In Russian).
121. Seymov, V. M. (1970). *Dynamic contact problems*. Kiev: Naukova Dumka. (In Russian).
122. Shcherban, N. G. (1986). *Building Materials, 12*, 13. (In Russian).
123. Shevtsov, V. R. (2010). *Strategy of re-renewal of resources from TPP ash-and-slag wastes of generating enterprises included in the Siberian Energy Association*. Novosibirsk: SEA. (In Russian).
124. Shpirt, M. Y. (1991). *Waste-free technology. Recycling of solid fuel extraction and processing wastes*. Subsoil: Moscow. (In Russian).
125. Shpirt, M. Y., Ruban, V. A., & Itkin, Y. V. (1990). *Rational use of coal mining and processing waste*. Moscow: Nedra. (In Russian).
126. Shvayka, D. I. (1990). *Handbook of the master in the production of wall ceramics*. Kiev: Budivelnik. (In Russian).
127. Sivertsev, G. N. (1961). *Chemical processes of hardening of concrete*. Moscow: Gosstroyizdat. (In Russian).
128. Slavutsky, A. K., Elenovich, A. S., Kurdenkov, B. I., & Romadanov, G. A. (1965). *Travel clothing made from local materials*. Moscow: Transport. (In Russian).

129. Smirnov, A. V. (1975). *Dynamics of road pavements of highways*. Omsk: West Siberian Publishers. (In Russian).
130. Smirnov, B. V., Trunov, B. D., Trufanov, V. N., & Shamray, I. A. (1972). *Lithological and petrographic features of heaving rocks of the East Donbass. Geological structure of the Rostov and adjacent areas*. Rostov-on-Don: Rostov State University Press. (In Russian).
131. Sneddon, I. N. (1951). *Fourier transforms*. New York: McGraw-Hill.
132. Solomatov, V. I. (1998). *Gorny Journal, 7–8*, 202. (In Russian).
133. Solomatov, V. I., Korenkova, S. F., & Chumachenko, N. G. (1999). *Building Materials, 7-8*, 12. (In Russian).
134. Solomatov, V. I., Korenkova, S. F., & Chumachenko, N. G. (1999). *Bulletin of Construction Equipment, 7-8*, 5. (In Russian).
135. Strokova, V. V. (2004). *Gorny Journal, 1*, 78. (In Russian).
136. Svatovskaya, L. B., Sheshneva, M. V., & Sycheva, A. M. (2006). *Ecology and Industry of Russia, 12*, 30. (In Russian).
137. Sychev, M. M., Kazan, E. N., & Petukhov, A. A. (1987). *News of Universities. Chemical Technology, 10*, 85. (In Russian).
138. Tanabe, K. (1971). *Solid acids and bases: Their catalytic properties*. London: Academic Press.
139. (1999). *Temporary method* of *determining* the *prevented environmental damage*. Moscow: Goskomekologiya. (In Russian).
140. (1986). *Temporary standard methodology for determining the economic efficiency of environmental protection measures and assessing the economic damage caused to the national economy by environmental pollution*. Moscow: Economics. (In Russian).
141. Trubetskoy, K. N., Rogov, E. I., & Nikitin, M. B. (1986). *The method of substantiating the optimal parameters for the creation and development of technogenic deposits. Integrated use of mineral raw materials*. Alma-Ata: Science. (In Russian).
142. Trubetskoy, K. N., Terpogosov, Z. A., & Shitarev, V. G. (1994). *Gorny Journal, 3*, 44. (In Russian).
143. Trubetskoy, K. N., & Umanets, V. N. (1992). *Gorny Journal, 1*, 12. (In Russian).
144. Trubetskoy, K. N., Umanets, V. N., & Nikitin, M. B. (1987). *Gorny Journal, 12*, 18. (In Russian).
145. Trubetskoy, K. N., Umanets, V. N., & Nikitin, M. B. (1989). *Gorny Journal, 12*, 6. (In Russian).
146. Trubetskoy, K. N., & Vorobiev, A. E. (1995). *Gorny Journal, 5*, 47. (In Russian).
147. Trunov, B. D. (1987). *Guidelines for the study and use in the national economy of waste dumps of coal mines in the Rostov region*. Rostov-on-Don. VNIGRIugol. (In Russian).
148. Tselykovsky, Y. K. (2002). *Engineering Ecology, 6*, 2. (In Russian).
149. Tselykovsky, Y. K. (2006). *Energetik, 10*, 29. (In Russian).
150. Tumanov, E. S., Tsibizov, A. N., & Bloch, N. T. (1991). *Technogenic resources of mineral raw materials*. Moscow: Nedra. (In Russian).
151. Ukhaneva, M. I., & Khobotova, E. B. (2010). *Bulletin of the Kharkiv National University, 895*, 260. (In Russian).
152. Umanets, V.N., Tolumbaev, A.Z., Kogut, A.V. (1990). In: *Abstr. of the All-Union Scientific and Practical Conference. Chelyabinsk*. (In Russian).
153. (1981). *Use of thermal energy and coal industry waste for the production of building materials in the COMECON countries (foreign experience)*. Express information VNIIESM, Ser. 20, 5, (In Russian).
154. Velichko, E. G., & Belyakova, J. S. (1966). *Building Materials, 5*, 27. (In Russian).
155. Vlasov, V. V. (1997). *News of Universities. Construction., 5*, 42. (In Russian).
156. Vlasova, V. V., Artemova, O. S., & Fomina, E. Y. (2017). *Ecology and Industry of Russia, 21*(11), 36. (In Russian).
157. Volzhensky, A. V., Burov, V. I., & Vinogradov, B. N. (1969). *Concrete and products from slag-ash materials*. Moscow: Stroyizdat. (In Russian).

158. Volzhensky, A. V., Ivanov, I. A., & Vinogradov, B. N. (1984). *The use of ash and fuel slag in the production of building materials*. Moscow: Stroyizdat. (In Russian).
159. Vorobiev, H. S. (1995). *Building Material, 10*, 6. (In Russian).
160. Vorobyov, A. V., Chekushina, T. V., & Vorobyov, S. E. (1990). *Ways and methods of forming technogenic mineral objects in the open development of complex structural deposits*. Moscow: TsNIIEItsvetmet. (In Russian).
161. Vorovich, I. I., & Babeshko, V. A. (1979). *Dynamic mixed problems of elasticity for non-classical domains*. Moscow: Nauka. (In Russian).
162. Worms, Y. N. (1994). *Building Materials and Structures, 1*, 4. (In Russian).
163. Xintaris, V. N., & Rekitar, J. A. (Eds.). (1983). *Use of secondary raw materials and waste in production (domestic and foreign experience, efficiency and trends)*. Moscow: Economics. (In Russian).
164. Yadykina, V. V. (2003). *News of Universities. Construction, 9*, 75. (In Russian).
165. Yadykina, V. V. (2004). *Science and Technology in the Road Industry, 1*, 45. (In Russian).
166. Yuriev, A. G., Lesovik, R. V., & Panchenko, L. A. (2006). *Concrete and Reinforced Concrete, 6*, 2. (In Russian).
167. Zalkind, I.Ya., Romanova, N.R., & Migachev, V.F. (1980). In *Physico-chemical fundamentals and environmental problems of the use of waste mining and processing of solid fossil fuels*. Part I, Moscow: Institute of Fossil Fuels of the USSR SA. (In Russian).
168. Zhukov, I. A. (2012). *TERRA ECONOMICUS, 10*(4), 123. (In Russian).
169. Zinchenko, V. N., & Pokutnev, Y. A. (1988). In *Utilization of industrial waste in the construction and repair of pavement in the south of the RSFSR*: Collection of works. Rostov-on-Don: Rostov Engineering-Building Institute Press. (In Russian).
170. Zyryanov, V. V. (2009). *Fly ash – technogenic raw materials*. Moscow: IIC "Mask". (In Russian).

Index

A

Abrasion, 12, 35, 44, 45, 81, 97, 108, 109, 111
Abrasive rock, 29
Access railway, 15, 78, 79
Accuracy criterion, 135
Acid oxides, 11, 47, 63
Activator additives, 73
Activity, 10, 11, 37, 38, 47, 57, 69–73, 75, 78, 86, 88, 104, 106, 109, 114–118, 123, 129, 130
Additive, 11, 12, 15, 17, 18, 25, 26, 38, 59, 64, 69–75, 77, 85, 86, 88, 89, 95, 96, 99, 104, 106, 108–110, 112, 114, 115, 117, 118, 122, 126
Adhesion, 34, 39, 80, 83, 93, 97–100, 110, 114–116
Adsorbent, 72
Adsorption activity, 73–75, 88, 118
Alkaline oxides, 35, 36
Aluminum, 11, 16, 44, 47, 48, 53, 54, 63, 68, 78, 115
Ambient temperature, 43
Amorphous components, 11, 66, 85
Amplitude displacement function, 189
Amplitude-frequency characteristic (AFC), 136, 146, 147, 150, 200
Amplitude-time characteristic (ATC), 135, 139, 142, 147–150, 154, 171, 193, 200
Analytical solution, 136, 145–150
Anomalous structure, 151, 152, 171–188
ANSYS, 138, 152, 190
Anthracite, 44–46, 49, 55, 59, 63, 65, 66
Argillite-siltstone compaction, 25
Ash-slag waste, 3–6, 10–14, 16–18, 30–39, 60, 62–65, 67–73, 84, 90, 95, 99, 105–110, 112–116, 122, 126, 129, 130
Asphalt concrete, 14, 16–18, 22, 25, 26, 34, 38, 91, 94, 97–103, 115–116, 192, 201–204, 221, 222
Atomization, 30

B

Ball mill, 26, 85
Basicity modulus, 47, 64
Batch mode, 191
Bending stiffness, 201
Bending strength, 95, 105, 108, 110
Binders, 10, 14, 16–18, 26, 38, 59, 69, 72, 75, 77, 85–88, 95–100, 103, 109, 110, 113, 115–117
Bitumen, 92, 94, 97–100, 115, 116
Black block, 43, 50
Boundary wave, 194
Brick, 25, 34, 43
Bronsted centers, 115, 116
Building materials, 10, 13, 15, 29, 33–39, 59, 60, 126–129
Bulging, 39, 80, 83
Bunker, 24, 26–28, 30, 33, 59, 60
Burning waste, 4
Burnt rocks, 6, 9–39, 41–75, 77–130
Bypass road, 15

A. A. Lyapin et al., *Improving Road Pavement Characteristics*, Innovation and Discovery in Russian Science and Engineering,
https://doi.org/10.1007/978-3-030-59230-1

C

Carbon, 4, 11, 30, 42, 44, 47, 50, 57, 63, 66, 68, 85
Carbonaceous matter, 11, 57, 58
Carbonaceous schists, 78, 90
Carbon-bearing rock, 5, 48
Cement stone, 10, 104, 107, 109, 117, 118
Ceramic, 22, 25, 26, 34, 45
Channel wave, 194
Chemical composition, 11, 45–47, 52, 59, 60, 63–65, 78, 84, 85, 97
Classifier, 22, 24, 27, 28, 32
Clay, 11, 12, 18, 21, 25, 31, 34–36, 42, 43, 45, 47–50, 52–58, 62, 66–72, 78, 81, 82, 84, 85, 89, 90, 97, 104, 110, 116, 117
Clay-glandular module, 63, 64, 97
Clayness, 47
Cleavage, 114
Clogging impurities, 34, 104, 116
Coal mining, 3, 4, 17, 43–45, 124
Coal seam, 5, 43, 69, 90
Coastal slope, 134, 137, 138, 142, 151
Coating, 14–16, 24, 25, 34, 37, 84, 89–94, 96–112, 190, 191, 194, 195, 198, 200–206, 211–214, 219–222
Coking products, 11, 35, 57
Combustion, 5, 30–32, 42, 43, 50, 59, 63, 66, 69, 70
Compactibility, 94
Compaction ratio, 39, 80, 83
Complex plane, 136
Compressibility, 15
Computer software, 138
Concrete, 10, 17, 24–26, 30, 34, 42, 75, 103–112, 116–122, 126, 127
Conditioned material, 23
Contaminating inclusion, 27
Corns-cuttings, 101
Corrosion resistance, 15, 109
Crushed stone, 14, 16, 20–23, 25–30, 33–39, 42, 80–85, 89–95, 97, 98, 100, 103, 109, 110, 114, 124–128
Crushing, 20, 22–30, 33–37, 81, 84–91, 97, 98, 103, 104, 109, 110, 124–128
Crystal lattice, 12, 68–70
Cup of instant deflections, 202
Cup of maximum deflections, 222

D

Damage, 14, 123–125, 190
Damping belt, 134, 136–137, 141–144, 189, 194, 195, 197, 200
Damping coefficient, 194
Delta function, 146
Density, 12, 13, 30, 31, 35, 36, 39, 44, 45, 59–62, 71, 73, 79–81, 83, 84, 93–97, 100, 101, 104, 107, 109, 114, 130, 138, 153, 172, 182, 190, 193, 194, 203, 204
Differential thermal analyzers, 42
Discretization, 195, 196
Dispersion, 12, 17, 37, 50, 54, 62, 71–73, 99, 110, 194
Displacement, 60, 133, 135, 136, 139–149, 154–164, 166–189, 198, 199, 202, 204, 221
Donor-acceptor mechanism, 116
Drainage layer, 84
Ductility, 12
Durability, 11, 20, 57, 64, 78, 84, 103, 107, 109, 112, 116, 118, 129
Dusty fractions, 4
Dye, 71–73, 88, 118
Dynamic deflection, 202, 204, 215–218, 221
Dynamic impact, 135, 144, 171, 222
Dynamic modeling, 133–150

E

Economic effect, 17, 124, 128
Economic indicators, 12–13
Electrostatic method, 16
Embankment, 13, 15, 18, 79, 82, 83, 89, 114
Emission, 10, 125
Endothermic effect, 51, 55, 120–122
Enrichment, 4, 6, 20–26, 31, 32, 44
Environmental indicators, 10
Environmental safety, 3, 4, 42
Erosion, 71, 73
Excitation, 145
Exothermic effect, 51, 56, 122

F

Fan, 26
Fatigue properties, 100
Feldspar, 47, 50, 52, 54, 58, 66–68
Fertile soil, 82
Fictitious boundary, 139, 195, 197
Fictitious wave, 135
Filler, 3, 13, 17, 18, 22, 26, 34, 37, 38, 59, 77, 84, 89, 90, 95, 97–100, 103, 107–110, 114, 116–118, 126
Filtration, 35, 36, 39, 80, 81, 83
Fines, 14, 16, 20, 23, 26–28, 32, 38, 44, 45, 62, 82, 93, 98, 99, 103, 108, 109, 114, 117, 118

Finite-element method (FEM), 133, 134, 136–138, 146, 149, 150, 152, 190–194, 222
Flaky grains, 103
Flotation, 31, 32
Fly ash, 17, 18, 30, 37, 39, 59, 61–63, 65, 67–69, 72–75, 82, 84–90, 95, 99, 103, 104, 109, 110, 114, 117, 118, 126–129
Fourier series, 146
Fourier transform, 145, 146, 149, 191
Fractionation, 21, 31, 32
Freezing, 14, 17, 31, 60, 95, 112
Frequency response, 136, 146–150, 171, 191, 192
Frequency spectrum, 146, 153, 190, 191
Frost resistance, 15, 35, 39, 45, 80–83, 88–90, 96, 97, 100, 103, 107–111, 116, 117
Fuel slag, 103

G

Gaseous compound, 4
Gel phase, 118
Geometric characteristics, 154, 171, 186, 196, 202, 204
Glandular, 12, 45, 53, 56, 70
Glass phase, 11, 30, 31, 67, 70, 71, 85, 115, 117, 118
Grade, 22–27, 35, 36, 39, 80–84, 88, 90, 94, 97, 98, 104, 107, 110–112, 120
Granulometric composition, 19, 37, 45, 60, 62, 91, 103, 114
Grates, 24
Gravel, 3, 20, 24, 31, 34, 35, 37, 78, 81, 90, 93, 94, 96–98, 110, 114, 126
Grinding, 24, 26, 45, 73–75, 85, 86, 88, 96
Ground fraction, 26

H

Hardening kinetics, 72, 86, 89
"Hard–soft–hard" structure, 151, 171–175, 178, 179, 181
Harmful substance, 10
Harmonic analysis, 145, 146, 189
Harmonic oscillations, 138, 145
Heat energy waste, 4, 32
Heavy metals, 10
Hematite, 11, 50, 53, 54, 66, 67
Heteropolar bond, 114
Histogram, 155, 156, 165, 186
Horizontal layer, 26
Humidity, 31, 38, 39, 43, 79, 80, 83, 86, 99, 104, 105, 114, 119, 122, 130
Hydration process, 72, 118
Hydraulic activity, 11, 14, 17, 30, 62, 69–75, 82, 88, 99
Hydrogarnet, 122
Hydrogenite, 70
Hydromica, 47, 48, 50–52, 54, 67
Hydrosilicate, 70, 71, 107, 110, 112, 115, 118, 120–122
Hygienic properties, 10

I

Ignition, 35, 36, 38, 43, 46, 47, 63, 64, 78, 85, 97
Improper contour integral, 136
Inclination angle, 152, 157, 158, 165, 173, 174, 182, 184, 186
Industrial goods, 9
Industrial waste, 3, 4, 6, 9, 10, 17, 78, 84, 86, 100, 126, 128
Inflammability, 43
Integral method, 145
Internal friction, 39, 80, 83
Internal wave, 194
Iron, 11, 43, 44, 47, 48, 50, 53, 54, 59, 60, 63, 66, 68, 115
Iron hydroxides, 11, 50, 53, 54, 58, 66
Isothermal heating, 104

K

Kaolinite, 47, 48, 50–52, 54, 68, 69

L

Lamé coefficients, 194
Lamellar grains, 27, 35, 81
Layered half-space, 134
Leading front, 158
Ledge, 26, 27
Lewis centers, 115, 116
Light microscopy, 42, 66
Limestone, 17, 48, 49, 62, 84–86
Limonite, 54
Linear algebraic equations (LAEs), 195, 198
Liquid glass, 95, 96
Load distribution, 138
Loam, 82
Longitudinal wave, 134, 138, 153, 172, 182, 195, 197
Loose placer, 93, 114
Low-frequency range, 148

M

Magnetite, 11, 50, 54, 66, 67
Marcasite, 47, 57, 66
Mathematical modeling, 5
Maximum permissible values (MPVs), 10
Mechanical characteristics, 45, 136, 142, 154–156, 164, 165, 171, 190, 191, 196, 197, 201, 202, 204
Metamorphism, 43, 44, 64
Metastable state, 11
Methylene blue, 73, 88, 118
Mica, 11, 47, 50–52, 58, 67
Mine dump, 4, 6, 9–39, 41–75, 77–130
Mineralogical composition, 5, 12, 43, 48, 58, 60, 63, 66, 72
Mineral powder, 17, 98–100
Mineral resources, 3, 5–7, 9
Mine rocks, 3, 5, 6, 10–15, 17, 18, 20, 23–27, 31, 34, 41, 43–45, 47, 50, 58, 69, 70, 73, 77–82, 85, 89, 90, 93, 97, 99, 102, 103, 105, 107, 113, 115, 118, 123–126
Modulus of elasticity, 96, 138, 204, 221
Moisture, 11, 30, 31, 35, 50, 51, 55, 60, 64, 78, 81, 95, 96, 104, 122
Mudstone, 47–53, 72, 78, 97

N

Natural radionuclides, 10, 37, 38, 78
Needle-like grains, 35, 114
Neoplasms, 66, 69, 70, 116, 118, 121
Nitrogen, 4
Non-stationary impact, 135, 146
Normalizing factor, 147, 148

O

Optical density, 73, 88, 118
Ore materials, 15
Oxidation, 11, 35, 43, 47, 50, 56, 57, 64, 66, 78
Oxide, 4, 11, 35, 36, 39, 43, 47, 48, 50, 59, 63, 64, 68, 71, 78, 84, 85, 115

P

Particle size distribution, 5, 31, 38, 60, 63, 79, 85, 94, 99, 114, 115
Pavement, 13, 14, 16–18, 23–25, 34, 39, 42, 64, 77, 84–97, 99–101, 103, 113–122, 130, 189–222
Paving slab, 17, 103, 106–109
Pebble, 78
Pelite, 53
Perturbation, 139, 148, 149
Petrographic characteristics, 11
Photocolorimetric method, 72
Physico-mechanical properties, 5, 12, 13, 20, 32, 34, 45, 52, 60, 79, 86, 89, 95, 96, 99, 100, 106–108, 111, 112, 116
Plane problem, 134, 136
Plasticity, 36, 39, 44, 80, 83, 110
Plasticizing additives, 99
Pneumatic method, 32
Point method, 42
Poisson's ratio, 138, 153, 172, 182, 203, 204
Pollution, 124, 125, 129
Porosity, 15, 35, 39, 45, 60, 62, 80, 83, 93, 97, 98, 100, 110, 117
Pozzolanic additives, 69, 71
Pozzolanic concrete, 18
Pozzolanic reaction, 71, 72
Pozzolans, 64, 69, 71, 117
Pulp, 31, 59, 60
Pyrite, 35, 36, 43, 47, 49, 57, 66

Q

Quarry, 14, 25, 84, 124, 125
Quartering method, 42
Quartz, 47, 48, 50–54, 56, 58, 63, 66–68, 70, 71

R

Railway tracks, 15, 78, 79, 82
Rayleigh waves, 171, 193
Reactivity, 43, 70, 73, 96, 104, 122
Recycling schemes, 19–33
Reflected wave, 135, 136, 139–142, 144, 149, 182, 194, 196
Refracted wave, 193
Re-melted rocks, 43, 50
Representative volume, 134–137, 139, 141, 148–150, 153, 194, 195, 197, 200
Residues, 5, 11, 35, 38, 39, 57, 64, 66, 80, 83
Rheological properties, 103
Road construction, 3, 4, 10, 13, 14, 16–18, 23, 30, 31, 37, 44, 59, 77–113, 123–130, 190–193, 197, 200–202, 204–222
"Road construction–soil" system, 190, 191, 194, 196–197, 203, 204, 222
Roller, 81, 82, 92–94, 96
Rotary crusher, 20, 27
Runways, 26

S

Sampling, 33, 41–42, 99
Sand, 14–18, 20, 21, 24–26, 28, 30, 31, 34, 35, 37, 38, 47–49, 58, 62, 66, 78, 82, 89–92, 95, 98–100, 103, 104, 109, 110, 126–128
Sandstone, 45, 48–50, 52, 53, 70, 78, 90, 97
Scanning electron microscopy, 52, 107
Screen, 24, 26–29, 32–34, 82, 171
Screening, 20–22, 25, 27–31, 33–37, 81, 84–89, 97, 98, 103–105, 109, 110, 124, 126–128
Sedimentation, 93
Selective extraction, 22
Semi-infinite layer, 151–188
Semi-limited structure, 134
Shear, 15, 84
Shear wave, 194
Siderite, 47, 51, 53
Sidewalk, 13, 15, 89, 90, 98, 99, 103, 107
Sieves, 28, 38, 39, 80, 83, 85
Silicate, 11, 47, 50, 66, 71, 72, 118
Silicon, 11, 44, 47, 48, 63, 78
Silicon dioxide, 11
Siltstone, 48–51, 53, 54, 58, 72, 78, 97
Silty, 21, 50, 97
Slag, 5, 6, 13, 16–18, 30–33, 37–39, 42, 44, 50, 59, 60, 62–70, 72, 73, 75, 82–85, 87–89, 95, 96, 100, 103, 104, 109, 110, 114, 118, 128
Slope, 14, 19, 26, 45, 82, 94, 137, 142–145, 152–180, 182–188
"Soft–hard–soft" structure, 151, 182, 184–186, 188
Soil, 13–15, 17, 18, 31, 39, 78–80, 82, 83, 89, 125, 129, 133–137, 143, 151, 152, 154, 171, 189–191, 194, 196, 197, 201–204, 221, 222
Soil base, 133–150
Solid mass, 133–150
"Solid mass–slope" system, 145
Sorption, 97
Specific surface, 38, 60–62, 70–75, 85, 110, 117
Spectral analysis, 10, 48, 64, 65
Spectral function, 192
Stepped surface, 152
Stiffness, 99, 151–188
Stone, 3, 37, 42, 75, 78, 82, 90, 98, 99, 106, 109, 117, 118
Strength, 5, 12, 14, 19–25, 27, 28, 30, 35, 36, 39, 44, 45, 71–73, 75, 80–84, 86–89, 94–98, 100–105, 107–118, 120, 122, 129, 201, 204, 221
Strengthening, 10, 16, 17, 82, 89, 104, 115, 221
Stress-strain state, 133–136, 139, 141, 142, 171, 191, 196, 200, 201, 204
Strip, 26, 190
Structural bond, 73, 78, 113, 117
Subgrade, 13–17, 64, 77–84, 89, 93, 113–114, 130, 190, 191, 194, 201–203, 209, 210, 221, 222
Sublattice, 20, 23
Subsoil, 3, 6, 128
Sulfonation, 43
Sulfur, 4, 11, 36, 43, 44, 47, 54, 57, 58, 63, 64
Superposition, 194
Supersaturation, 118
Surface energy, 114
Surface wave, 171, 182, 193, 194
Synergistic effect, 75, 116

T

Technical indicators, 12–13
Technogenic deposits, 4–6, 12
Technogenic raw materials, 3–7, 9–13, 17, 42, 44, 73, 77, 79, 84–90, 97–116, 118, 129, 130
Technological scheme, 13, 20–26, 28, 29, 33, 34, 42
Thawing, 95, 112
Thermal insulation properties, 14, 17, 97
Thermal power plant (TPP), 62–64, 82–84, 95, 100
Third-order matrix function, 138
Three-layer structure, 137, 138, 152, 153, 172, 182
Tiny spherulite, 103
Toxic substances, 10
Transmission electron microscopy, 52
Transverse wave, 135, 138, 153, 172, 182
Triangular pulse, 192
Triple diagram, 65

U

Ultimate compressive strength, 45
Unburnt rocks, 5
Underlying layer, 82, 84, 89, 91, 93, 94
Utilization, 4, 6, 12, 13, 16–18, 32, 52, 107, 117, 123–125, 129

V

Vibrating screen, 22, 24, 26
Viscosity, 99, 134, 136, 142, 190, 195, 197
Vitreous shell, 73, 85
Voidness, 45, 103, 104

W

Warehouse, 23, 24, 30, 32, 34, 60, 107
Washing, 21, 29, 30
Waste dump, 5, 26, 33, 78, 90, 124
Water absorption, 35, 45, 80, 81, 88, 97, 107, 108, 111, 117
Water demand, 62, 103, 104
Water resistance, 12, 35, 39, 71, 72, 80, 81, 83, 90, 96, 97, 100–102, 107, 109, 110, 116, 117
Water-saturated state, 97
Wave packet, 155–157, 159–162, 164, 166–169, 172–178, 180, 182–185, 187
Wave propagation velocity, 134, 138, 153, 172, 182
Weak differences, 20–22, 24
Wear resistance, 103
Weathering, 47, 50, 54, 66, 90
Workability, 104, 110

X

X-ray diffraction, 42, 52, 54, 57